Swen König

Semiconductor Optical Amplifiers and mm-Wave Wireless Links for Converged Access Networks

Karlsruhe Series in Photonics & Communications, Vol. 14
Edited by Profs. J. Leuthold, W. Freude and C. Koos

Karlsruhe Institute of Technology (KIT)
Institute of Photonics and Quantum Electronics (IPQ)
Germany

Semiconductor Optical Amplifiers and mm-Wave Wireless Links for Converged Access Networks

by
Swen König

KIT Scientific Publishing

Dissertation, Karlsruher Institut für Technologie (KIT)
Fakultät für Elektrotechnik und Informationstechnik, 2013

Impressum

KIT Scientific Publishing

Karlsruher Institut für Technologie (KIT)
KIT Scientific Publishing
Straße am Forum 2
D-76131 Karlsruhe

KIT Scientific Publishing is a registered trademark of Karlsruhe Institute of Technology. Reprint using the book cover is not allowed.

www.ksp.kit.edu

Print on Demand 2014

ISSN 1865-1100
ISBN 978-3-7315-0255-5

Semiconductor Optical Amplifiers and mm-Wave Wireless Links for Converged Access Networks

Zur Erlangung des akademischen Grades eines

DOKTOR-INGENIEURS

von der Fakultät für
Elektrotechnik und Informationstechnik
des Karlsruher Instituts für Technologie (KIT)

genehmigte

DISSERTATION

von

Dipl.-Ing. Swen König

geboren in: Schwäbisch Gmünd

Tag der mündlichen Prüfung: 08.07.2013
Hauptreferent: Prof. Dr. sc. nat. Jürg Leuthold
Korreferenten: Prof. Dr.-Ing. Dr. h. c. Wolfgang Freude
Prof. Dr.-Ing. Ingmar Kallfass

„Damit das Mögliche entsteht, muss immer
wieder das Unmögliche versucht werden."

– Hermann Hesse (1877-1962)

Table of Contents

Abstract (German)

Die vorliegende Arbeit befasst sich mit der Verwendung von optischen Halbleiterverstärkern (*engl.*: semiconductor optical amplifier, SOA) in zukünftigen konvergierten Zugangsnetzen (*engl.*: converged access network) sowie mit der Einbindung von hochbitratigen Millimeterwellen-Funkverbindungen (*engl.*: mm-wave wireless link) in derartige Netze. In Systemexperimenten werden sowohl das Potential als auch die Limitierungen dieser beiden Netz-Elemente aufgezeigt und darüber hinaus mehrere praxisrelevante Anwendungsszenarien erstmalig demonstriert.

Der stark anwachsende globale Internet-Datenverkehr, neue informationstechnische Anwendungen und Dienste wie z.B. die „Informationsverarbeitung in der Wolke" (*engl.*: cloud computing) oder das Betrachten hochaufgelöster Videoinhalte über das Internet sowie das Bedürfnis des Nutzers, möglichst überall und jederzeit über einen breitbandigen Zugang zu Daten und Diensten zu verfügen, stellt eine große Herausforderung für die Kommunikationsinfrastruktur dar, die entsprechend diesen neuen Anforderungen angepasst und ausgebaut werden muss. Dies betrifft sowohl die kabelgebundenen als auch die drahtlosen Kommunikationsnetze. Aufgrund des mittlerweile enormen Bandbreitenbedarfs sind beide besonders auf den intelligenten Ausbau optischer Fasernetze angewiesen. Beispielsweise sind die Datenaufkommen in den Mobilfunknetzen der vierten Generation (LTE, *engl.*: long term evolution) derart groß, dass einzig eine direkte Anbindung der LTE-Zugangspunkte an eine optische Faserinfrastruktur die benötigte Bandbreite für den Weitertransport der Daten bietet.

Konvergierte Netze werden eine wichtige Rolle in zukünftigen Zugangsnetzen spielen. Die Untersuchungen im Rahmen der vorliegenden Arbeit liefern Erkenntnisse im Zusammenhang mit dem Zusammenwachsen (der „Konvergenz") von faserbasierten optischen Zugangsnetzen und Funknetzen sowie der Konvergenz von optischen Zugangs- und Metronetzen. Bei der Konvergenz von optischen Zugangsnetzen und Funknetzen werden breitbandige optische Glasfasern zur Übertragung von Daten genutzt, die aus drahtlosen Netzwerken kommen bzw. dafür bestimmt sind. Im einfachsten Fall werden die Funksignale direkt auf einen optischen Träger aufmoduliert (*engl.*: radio over fiber, RoF). Unter der Konvergenz von optischen Zugangs- und optischen Metronetzen versteht man dann das Zusammenlegen unterschiedlicher physikalischer Netze zu einer gemeinsamen Infrastruktur, welche gleichzeitig als Zugangsnetz und als Rückgrat (*engl.*: backbone) fungiert.

Optische Halbleiterverstärker gelten aufgrund ihres breiten und flexibel einstellbaren Gewinnspektrums, ihres kleinen Formfaktors und ihres günstigen Preises als vielversprechend für den Einsatz als Verstärker zur Verlustkompensation und somit zur Vergrößerung der Übertragungsdistanzen in zukünftigen konvergierten Zugangsnetzen. Dafür müssen die Verstärker mit den unterschiedlichsten Signalen und Multiplexverfahren zurechtkommen, ohne die Signalqualität entscheidend zu verschlechtern.

In Kapitel 3 wird die Verwendung von optischen Halbleiterverstärkern mit verschiedenen Signalen und Multiplexverfahren im Detail untersucht. Zunächst werden optische Inphasen-Quadratur-Signale (IQ) mit den Modulationsformaten BPSK, QPSK, 8PSK und 16QAM und jeweils einer Symbolrate von 28 GBd verstärkt. Für jedes Format wird der Eingangsdynamikbereich (*engl.*: input power dynamic range, IPDR) des Verstärkers bestimmt, für den eine Bitfehlerwahrscheinlichkeit (*engl.*: bit error ratio, BER) von höchstens 10^{-3} erreicht wird. Der Eingangsdynamikbereich des Verstärkers ist eine wichtige Kenngröße, da je nach Position des Verstärkers im optischen Netzwerk die Verstärker-Eingangsleistung sehr unterschiedlich sein kann. Das Verstärkerrauschen limitiert die Signalqualität bei kleinen Eingangsleistungen. Nichtlineare Effekte, Amplitudenkompression und Amplituden-Phasen-Kopplung im Verstärker degradieren die Signale bei großen Eingangsleistungen im nichtlinearen Verstärkungsbereich. Für die Formate BPSK und QPSK wird ein Eingangsdynamikbereich von über 40 dB gemessen, für 8PSK und 16QAM beträgt er noch 29 dB bzw. 26 dB. Es wird außerdem der Eingangsdynamikbereich für IQ-Signale untersucht, die in einem Wellenlängen- und/oder Polarisations-Multiplex-Verfahren kombiniert und dann verstärkt werden.

Momentan gibt es Überlegungen, in ringförmig aufgebauten Metro-Zugangsnetzwerken optische Halbleiterverstärker in rekonfigurierbaren optischen Add-Drop-Multiplexern (*engl.*: reconfigurable optical add-drop multiplexer, ROADM) in Abständen von 5…20 km einzusetzen. Die Datensignale auf dem Ring durchlaufen mehrere Add-Drop-Multiplexer und somit eine Verstärkerkaskade. In der vorliegenden Arbeit wird eine Kaskade aus 4 Halbleiterverstärkern untersucht, welche optische Ein- und Mehrkanal-Signale verstärkt. Der Eingangsdynamikbereich der Kaskade beträgt nach dem vierten Verstärker für Einkanal-Signale mit einer Symbolrate von 28 GBd und den Formaten BPSK, QPSK, 8PSK und 16QAM noch 23 dB, 20 dB, 12 dB und 9 dB. Entlang der Kaskade wird zudem der Eingangsdynamikbereich für optische IQ-Signale bestimmt, die im Wellenlängen- und/oder Polarisations-Multiplex-Verfahren übertragen werden.

Im Hinblick auf die Konvergenz von optischen Zugangsnetzen und Funknetzen wird in Kapitel 3 die Verstärkung elektrischer Hochfrequenz-Signale (HF) mit optischen Halbleiterverstärkern untersucht. Dabei wird das HF-Signal mittels Intensitätsmodulation auf den optischen Träger moduliert. Die Modulationstiefe ist so gering, dass es sich um eine Kleinsignalmodulation handelt. Im linearen Verstärkungsbereich ist der Betrag der Verstärkung unabhängig von der Frequenz des HF-Signals. Wird der Verstärker in Sättigung betrieben, dann wird aufgrund der Gewinndynamik ein Hochpassverhalten beobachtet.

Für HF-Signale wird neben dem optischen auch der elektrische Eingangsdynamikbereich des Halbleiterverstärkers (*engl.*: electrical power dynamic range, EPDR) bestimmt, welcher angibt, in welchem Bereich der Modulationstiefe die Bitfehlerwahrscheinlichkeit des HF-Signals am Ausgang des Verstärkers unterhalb einer definierten Grenze liegt. Es zeigt sich, dass der optische und elektrische Eingangsdynamikbereich voneinander abhängen.

In einem ausführlichen Systemexperiment wird die Beeinflussung eines optisch übertragenen HF-Signals durch optische Datensignale sowie durch ein zweites optisch übertragenes HF-Signal untersucht, wenn alle Signale gleichzeitig vom selben optischen Halbleiterverstärker verstärkt werden. Es wird die Verstärkung des Rückkanals eines optischen Zugangsnetzwerkes betrachtet, in dem Daten in einem kombinierten Wellenlängen- und Zeit-Multiplex-Verfahren übertragen werden.

Schließlich wird die Verwendung von optischen Halbleiterverstärkern zur Verstärkung breitbandiger, elektrischer OFDM-Signale untersucht. Die OFDM-Signale werden zudem über eine Kaskade von bis zu acht Halbleiterverstärkern übertragen. Wie erwartet werden in der Sättigung die OFDM-Subträger mit niedriger Frequenz durch das Hochpassverhalten des Verstärkers beeinflusst. Zudem verschlechtern Intermodulationsprodukte der Subträger untereinander die Signalqualität.

Derzeit findet der Ausbau faseroptischer Netzwerke vorwiegend in Ballungsräumen und dicht besiedelten Gebieten statt. In dünn besiedelten Gegenden wie beispielsweise im ländlichen Raum oder schwer zugänglichem Terrain verhindern die hohen Investitionskosten für eine neue Faserinfrastruktur die Versorgung mit einem schnellen Internetzugang.

In derartigen Fällen kann eine hochbitratige Millimeterwellen-Funkverbindung einen kostengünstigen und vollwertigen Ersatz zur optischen Faser darstellen. Dabei schließt die Funkverbindung nahtlos an eine bereits bestehende Faserinfrastruktur an und überträgt dann die Daten in diejenigen Versorgungsgebiete, wo keine optische Faser zur Verfügung steht. Im Frequenzbereich zwischen 200 und 300 GHz befindet sich ein atmosphärisches Fenster mit geringen Verlusten und ausreichender Bandbreite für eine Funkverbindung, mit der vergleichbare Datenmengen wie in optischen Zugangsnetzen übertragen werden können. Die Millimeterwellen-Signale werden entweder mit breitbandiger HF-Elektronik oder mit optischen Mitteln erzeugt. Für den Empfang gibt es bislang keine optische Alternative.

In Kapitel 4 wird erstmalig eine Millimeterwellen-Funkverbindung gezeigt, die bei einer Trägerfrequenz von 220 GHz operiert und nahtlos in ein optisches Umfeld eingebettet ist. Die Funkverbindung stellt eine drahtlose Verbindung zwischen dem End- und Anfangspunkt zweier optischer Faserstrecken her, die jeweils 20 km lang sind. Nach der optischen Übertragung über die erste Faserstrecke werden die übertragenen Daten entweder im Ein-Aus-Format (*engl.*: on-off-keying, OOK) mit einer Datenrate von bis zu 20 GBit/s oder die elektrischen OFDM Signale mit einer Datenrate von bis zu 9 Gbit/s mit einer Photodiode empfangen. Das im Photostrom empfangene Basisband wird mit einem elektronischen Mischer, der mit monolithisch integrierten Millimeterwellen-Schaltkreisen (*engl.*: mm-wave monolithic integrated circuits, MMIC) aufgebaut ist, auf die Trägerfrequenz von 220 GHz umgesetzt. Nach der Funkübertragung wird das Millimeterwellen-Signal mit einem MMIC-Mischer wieder ins Basisband umgesetzt, danach auf einen optischen Träger kodiert und über die zweite Faserstrecke übertragen.

In einem weiteren Experiment wird die bislang größte Datenrate von 100 Gbit/s mit einer SISO-Millimeterwellen-Funkverbindung (*engl.*: single-input single-output) gezeigt. Ermöglicht wird diese außerordentlich hohe Datenrate durch die Verwendung von spektral effizienten IQ-Signalen, einer breitbandigen Photodiode (PD, aufgebaut als „uni-travelling carrier photodiode“, UTC-PD), in welcher das Millimeterwellen-Signal mit einer Trägerfrequenz von 237,5 GHz durch Photomischung erzeugt wird, und eines breitbandigen MMIC-IQ-Mischers am Empfänger, der das über 40 m übertragene Signal wieder ins Basisband umsetzt. Für die Erzeugung des Millimeterwellen-Signals mittels Photomischung werden ausgewählte Kammlinien eines modengekoppelten Lasers (der Abstand benachbarter Kammlinien beträgt 12,5 GHz) verwendet. Entweder auf eine Spektrallinie (Einkanal-Signal) bzw. auf drei Linien (Mehrkanal-Signal) werden Daten kodiert. Eine weitere, unmodulierte Linie im Abstand von 237,5 GHz zur zentralen modulierten Linie dient als Lokaloszillator bei der Photomischung an der UTC-PD. Es werden QPSK-, 8PSK-, 8QAM- und 16QAM-Signale übertragen. Sowohl im Ein- als auch im Mehrkanal-Experiment wird eine Datenrate von 100 Gbit/s erreicht. Im Mehrkanal-Experiment werden die zu übermittelnden Signalimpulse der einzelnen Subkanäle mit einem Filter geformt, das ein cos-förmiges Spektrum aufweist (*engl.*: „raised-cosine“).

Preface

The exponential growth of data traffic in optical and wireless communication networks constantly poses new challenges to the underlying network infrastructures. Sometimes, small modifications of the existing infrastructure are sufficient to satisfy the customers' need for the near future. For a long-term perspective, however, versatile and scalable access network infrastructures are needed that can cope with the requirements of tomorrow's services to be offered in these networks. In this case, fundamental changes become necessary: On the one hand, the use of new physical-layer devices and innovative technologies are required. On the other hand, the network infrastructure itself needs to undergo evolution or even revolution.

The access network of the future is a converged network. Network convergence directly results from the customers' demands and needs, which today call for ubiquitous access to voice, data, video, and interactive multimedia services at any time and any place. The access network of the future must support all of these services, provide sufficient capacity for a large number of customers, and have a high-capacity infrastructure even for the "last mile" to the customer. In this thesis we discuss converged optical-wireless networks and converged metro-access networks.

Optical-wireless convergence means sharing the same infrastructure for fixed-line (copper or fiber) and wireless services. For instance, the fourth generation mobile communication networks (4G), also known as long term evolution (LTE) network, provide broadband wireless access solutions that offer mobility to the end-user. However, the large amount of mobile traffic generated in LTE networks must also be supported by a mobile backhaul network. For LTE, only a fiber infrastructure enables a suitable backhaul solution which offers sufficient bandwidth. Optical-wireless convergence also relies on so called radio-over-fiber (RoF) systems where radio signals are converted to an optical signal for transmission over an optical fiber network. These systems combine the advantages of huge bandwidth and low transmission losses in fiber-optic networks with the ubiquity and mobility of wireless access networks.

In a converged metro-access network infrastructure large parts of the metro network and of the access network are merged into a common infrastructure, e.g., a metro-access network usually with a ring topology. This network approach offers a large fiber-based capacity in a large coverage area. So-called reconfigurable optical add-drop multiplexers (ROADM) along the metro-access ring provide reconfigurable and/or fixed add/drop ports where either active system equipment or other infrastructure segments can be connected. Examples are a wavelength division multiplexed (WDM) passive optical network (PON), a fiber link for mobile backhauling (RoF), or a direct fiber connection to enterprise customers, data centers, and large institutions.

In this thesis we study the use of two physical layer elements in converged access networks: semiconductor optical amplifiers (SOA) and mm-wave wireless links.

Semiconductor optical amplifiers are considered to be key devices in future optical access networks. Semiconductor optical amplifiers have a small form factor, are cost-efficient and they offer a broad gain bandwidth of up to 100 nm that may be centered at virtually any desired optical communication wavelength. By using SOAs as loss compensating elements, the reach in access networks can be considerably extended. This in turn allows the realization of new access network architectures. In converged access networks, SOAs have to cope with different types of data signals, modulation formats, and multiplexing techniques. These questions are addressed in this thesis.

In cases where the deployment of new fiber infrastructure is not economical, a high-capacity mm-wave wireless link could be an adequate fiber substitute. If the wireless link is integrated into an optical fiber infrastructure, the wireless link virtually acts as a reach extender of the optical network ("fiber over radio"). Such a high-capacity wireless link can also be used for bridging difficult-to-access terrains, for radio backhauling, and as an *ad hoc* solution if an already existing fiber connection fails. In the frequency range of 200…300 GHz there is a low-loss atmospheric transmission window which offers sufficient bandwidth for a mm-wave wireless link that supports data rates comparable to those found in optical access networks. As another topic in this thesis, we investigate the integration of a high-capacity mm-wave wireless link operating at carrier frequencies of 220 GHz and 237.5 GHz into a fiber-optic environment.

In ***Chapter 1*** of this thesis, we motivate the use of SOAs and high-capacity mm-wave wireless links in the framework of converged access networks. Existing optical and mobile networks are briefly reviewed.

In ***Chapter 2***, fundamentals are presented, which are helpful for the understanding of the following material. We briefly discuss inphase-quadrature (IQ) modulation, typical transmitter and receiver implementations, Nyquist pulse shaping, orthogonal frequency division multiplexing (OFDM), SOA basics, and mm-wave electronics and photonics.

In ***Chapter 3***, we present the results of our investigations on SOAs in converged access networks. We discuss the use of SOAs with different types of data signals and multiplexing formats. In this context we study the amplification of single wavelength IQ signals employing 28 GBd BPSK, QPSK, 8PSK, or 16QAM formats. For each modulation format we determine the SOA input power dynamic range (IPDR), within which the bit error ratio (BER) is 10^{-3} or smaller. The IPDR is an important parameter. Depending on the position of the SOA in the network, the SOA input power may vary. For low input powers, the signal quality is degraded due to optical amplifier noise. For large input power levels, SOA nonlinearities, amplitude compression, and amplitude-phase coupling worsen the signal quality. For BPSK and QPSK, the IPDR is larger than 40 dB. For 8PSK and 16QAM the IPDR is 29 dB and 26 dB, respectively. Moreover, we determine the IPDR for multi-wavelength operation and/or polarization multiplex signals.

We then study a converged metro-access ring network with a cascade of four identical SOAs. The SOAs represent loss compensating elements in the ROADMs along the ring. Single-wavelength, multi-wavelength, and polarization multiplexed IQ signals are amplified by the SOA cascade, and the IPDR is determined after each SOA. For the 28 GBd single-wavelength BPSK, QPSK, 8PSK, and 16QAM signals, the IPDR after the fourth SOA is 23 dB, 20 dB, 12 dB, and 9 dB, respectively.

In the context of optical-wireless convergence, we investigate the amplification of RoF signals with an SOA. Radio signals are encoded on an optical carrier by intensity modulation. The optical modulation depth is small. A small-signal analysis of the SOA shows that in the linear operating region the SOA modulation response does not depend on the frequency of the radio signal. In saturation, a high-pass characteristic is observed due to the gain saturation dynamics. Besides the IPDR, we also measure the electrical power dynamic range (EPDR). The EPDR gives the range of optical modulation depth within which the BER of the radio signal after amplification is below a limiting value. We show that EPDR and IPDR are interdependent.

In a system experiment we study the SOA as a reach extender. We investigate the impact of a rival RoF signal, a rival WDM signal and a rival bursty time division multiplexed (TDM) signal on a RoF subscriber in a converged RoF and WDM-TDM-gigabit passive optical network (GPON) upstream scenario. The rival signal and the RoF signal are simultaneously amplified by the same SOA.

We also investigate the amplification of broadband electrical OFDM signals with an SOA. Moreover, the OFDM signals are transmitted over a cascade of up to 8 SOAs. As is expected, in saturation, the lower frequency OFDM subcarriers are influenced by the SOA high-pass characteristic when operating the SOA in the saturation regime. Furthermore, for large SOA input powers nonlinear interactions of the OFDM subcarriers (inter-subcarrier mixing) limit the performance.

In ***Chapter 4***, we demonstrate for the first time the feasibility of a wireless link at 220 GHz based on electronic up-conversion and down-conversion. The wireless link connects two optical fiber links at data rates of up 20 Gbit/s. Either non-return-to-zero (NRZ) on-off keying (OOK) with data rates up to 20 Gbit/s, or electrical OFDM with data rates up to 9 Gbit/s are used. The wireless link connects two spatially separated fiber sections, each with a length of up to 20 km.

Furthermore, we demonstrate for the first time single-input single-output (SISO) mm-wave transmission at 237.5 GHz with a record data rate of 100 Gbit/s. The mm-wave signal is generated photonically by heterodyning selected optical lines of a mode-locked laser (MLL) on a uni-travelling carrier (UTC) photodiode (PD). We demonstrate the generation of a SISO 100 Gbit/s single-channel signal using 16QAM. Moreover, we show the generation of a SISO 100 Gbit/s multi-channel signal by employing different modulation formats and sinc-like pulse shaping using three adjacent MLL lines. The data are successfully transmitted over a wireless distances of up to 40 m.

Achievements of the Present Work

In this thesis, semiconductor optical amplifiers (SOA) and high-capacity mm-wave wireless links are investigated for the use in converged access networks of the future. Special emphasis has been put on the applicability of SOAs and mm-wave wireless links in converged metro-access networks and in converged optical-wireless networks. The full potential and the limitations of the studied SOAs and mm-wave links have been identified and demonstrated in numerous characterization and proof-of-principle experiments.

In the following, we give a concise overview of the main achievements of this thesis.

Amplification of optical single-wavelength IQ signals:
The SOA input power dynamic range (IPDR) is determined when amplifying 28 GBd single-wavelength signals employing BPSK, QPSK, 8PSK or 16QAM formats. The limitations for small and large SOA input powers are studied, and the contribution of phase errors and magnitude errors to the error vector magnitude (EVM) is identified for the different investigated modulation formats.

Amplification of optical multi-wavelength IQ signals:
Optical multi-wavelength IQ signals with a symbol rate of 28 GBd per channel are amplified with an SOA. The obtained IPDR values are compared to the IPDR results for the amplification of single-wavelength IQ signals.

Amplification of optical polarization multiplexed IQ signals:
Optical single-wavelength and multi-wavelength IQ signals employing polarization multiplexing are amplified with an SOA. The influence of the polarization dependent SOA gain (PDG) on the IPDR is studied for QPSK and 16QAM formats.

Measured and calculated BER for linear and nonlinear amplification:
For 28 GBd QPSK and 16QAM signals, the directly measured bit error ratio (BER) and the calculated BER from EVM measurements are compared in the linear and nonlinear SOA operating regime. In the linear operating regime, the directly measured and the calculated BER from EVM measurements coincide. In the nonlinear operating regime, the directly measured BER is much larger than the calculated BER.

Amplification of radio-over-fiber (RoF) signals:

The optical small-signal response of an SOA is measured and described theoretically. In the nonlinear operating regime, the SOA exhibits a high-pass characteristic. With an SOA we amplify intensity-modulated RoF signals, which have an optical modulation depth between 1 % and 35 %. The influence of the SOA high-pass characteristic is compared for RoF signals with a carrier frequency of 2 GHz and 5 GHz, respectively. We also study the impact of the SOA high-pass characteristic on the different subcarriers of an ultra-broadband electrical orthogonal frequency division multiplexed (OFDM) signal occupying the frequency range 0…5 GHz. Besides the IPDR, we determine the electrical power dynamic range (EPDR) of an SOA when amplifying RoF signals.

Simultaneous amplification of different data types and signals:

We perform experiments where different kinds of data and multiplexed signals are simultaneously amplified by the same SOA. For the first time, we determine the impact of a rival WDM signal, a rival bursty time division multiplexed (TDM) signal, and a rival RoF signal on the upstream data of a RoF subscriber. Based on our experimental results, optimum operating conditions for the SOA are identified.

Concatenation of SOAs:

For the first time, a concatenation of SOAs for amplifying optical IQ signals is investigated. Single-wavelength, multi-wavelength, and polarization multiplexed signals employing BPSK, QPSK, 8QPSK, and 16QAM formats are amplified by an SOA cascade consisting of 4 identical SOAs. For the different modulation formats the IPDR is evaluated after each SOA. For QPSK modulation, the influence of the optical filter bandwidth on the IPDR along the SOA cascade is studied. The feasibility of high-capacity business paths with data rates of up to 448 Gbit/s is experimentally demonstrated for a metro-access ring network scenario, where SOAs are located in the reconfigurable optical add-drop multiplexers (ROADM) along the ring.

For the first time, we perform experiments using a recirculating fiber-loop with one QD SOA to investigate an SOA cascade with intensity-modulated and direct-detected multi-gigabit RoF OFDM signals. From our experimental results we identify the optimum SOA input power and the optical power dynamic range along the SOA cascade.

Integration of a 220 GHz wireless link into a fiber environment:

We demonstrate for the first time the feasibility of a wireless link at 220 GHz based on electronic up-conversion and down-conversion, thereby connecting two optical fiber links at data rates of up 20 Gbit/s. Either non-return-to-zero (NRZ) on-off keying (OOK) with data rates up to 20 Gbit/s, or electrical OFDM with data rates up to 9 Gbit/s are used. The wireless link connects two spatially separated fiber sections. Each fiber section has a length of up to 20 km.

100 Gbit/s SISO wireless link with mm-wave photonics:

We demonstrate for the first time a single-input single-output (SISO) mm-wave transmission at 237.5 GHz with a record data rate of 100 Gbit/s. So far, this is the largest data rate transmitted over a SISO wireless link. The mm-wave signal is generated photonically by heterodyning selected lines of a mode-locked laser (MLL) on a uni-travelling carrier (UTC) photodiode (PD). We demonstrate the generation of a SISO 100 Gbit/s single-channel signal using 16QAM. Moreover, we show the generation of a SISO 100 Gbit/s multi-channel signal by employing different modulation formats and sinc-like pulse shaping on three adjacent MLL lines. We successfully transmit the data over wireless distances of up to 40 m.

1 Introduction

At home or on the move – communication technology has become an essential part of our daily lives. In addition to well-established applications such as surfing the internet, emailing, online shopping, or social networking, a wealth of upcoming services like high definition television (HDTV), 3D television, video on demand (VoD) and cloud computing will significantly increase the amount of both fixed and mobile data traffic. Fixed data traffic includes data traffic from wired connections but also from stationary wireless links connecting two fixed sites. Mobile data traffic denotes the traffic generated by mobile devices, e.g., handsets, smartphones, and mobile broadband gateways.

An ongoing exponential growth of data traffic is predicted in the latest forecasts by Cisco Systems. According to [1] the annual global internet protocol (IP) traffic will reach 1.1 zettabytes (ZB) (or 110 exabytes (EB) per month) by the end of 2016, see Fig. 1.1(a). It is expected that the global IP traffic will grow at a compound annual growth rate (CAGR) of 29 %. Fixed wired and wireless IP traffic contributes the major part to the overall data traffic followed by managed IP traffic which includes IP transport of TV and VoD signals. While in 2011 the global average fixed broadband speed was 9.1 Mbit/s, a speed of 34.5 Mbit/s is projected for the year 2016.

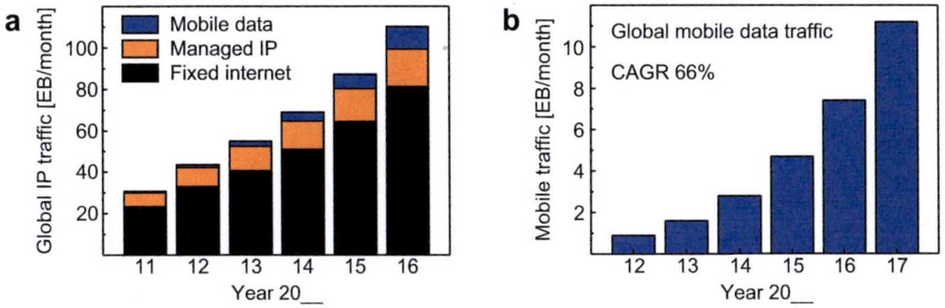

Fig. 1.1: In the next few years, the global IP traffic (a) is expected to grow with a compound annual growth rate (CAGR) of 29 %. The global IP traffic will reach 110 exabytes (EB) per month in 2016. The total IP traffic includes fixed (wired and wireless) internet data, managed IP data and mobile data. The global mobile data traffic (b) grows at a CAGR of 66 % and will reach 11.2 EB per month in 2017. This requires broadband access networks for fixed and mobile services.

Compared to the fixed IP traffic, only a small amount of data traffic seems to be generated by mobile data. However, the global mobile data traffic grows about three times faster than the fixed IP traffic. As stated in [2], the global mobile data traffic reached 885 petabytes (PB) per month at the end of 2012, see Fig. 1.1(b). This is nearly twelve times larger than the global IP traffic in 2000 (75 PB). A 13-fold increase of the global mobile data traffic is expected between 2012 and 2017, reaching 11.2 EB per month by 2017 which corresponds to a CAGR of 66%. In the same period, the average mobile connection speeds are expected to increase from 526 kbit/s (in 2012) to 3.9 Mbit/s (in 2017).

That said, it is evident that both the wired and the mobile broadband access space will undergo radical change. Novel versatile and future-proof access network topologies and infrastructures will evolve to satisfy the bandwidth requirements for tomorrow's services offered in these networks. The access network of the future will be a "converged" network and provide the customer with ubiquitous high-speed connectivity. The term "converged network" refers to convergence with respect to manifold aspects as will be explained in more detail in *Section 1.2* of this *Chapter*.

Fundamental changes of network infrastructures and new access network concepts require modifications, new devices, and new technologies already on the lowest networking level, the physical layer. In this work, two promising elements for the physical layer are investigated in the context of converged access networks.

The first element is a semiconductor optical amplifier (SOA). SOAs are considered to be key devices in future optical access networks. SOAs have a small form factor, are cost-efficient and offer a broad gain spectrum that may be centered at virtually any desired optical communication wavelength in the spectral range of 1250 nm to 1600 nm. By using SOAs, the reach in access networks can be considerably extended. This in turn allows the realization of new access network architectures.

The second element is a high-capacity mm-wave wireless link. Today, the radiofrequency spectral resources are extremely limited because of the heavy use of today's conventional frequency range up to 60 GHz. The unlicensed mm-wave band around 60 GHz offers a bandwidth of 7 GHz and already enables multi-gigabit wireless data transmission in the order of 10 Gbit/s. However, a further significant capacity enhancement requires even larger bandwidths, which are only available in the high mm-wave and THz region. Between 200 GHz and 300 GHz, there is a transmission window with low atmospheric losses where wireless links can be operated even under difficult weather conditions like heavy rain or fog. The mm-wave wireless link presented in this thesis operates at a carrier frequency of 237.5 GHz and supports a data rate of 100 Gbit/s. This data rate is comparable to the data rates in next generation optical acccss networks which makes it attractive to integrate such a wireless link directly into the access network infrastructure. There it can serve as an adequate substitute for an optical fiber connection, for instance if an end-to-end fiber connection is absent or damaged. In this sense, the mm-wave link acts as a wireless reach extender in optical network.

1.1 Review of Existing Optical and Mobile Networks

The content of this *Section* is considered as standard knowledge. Major parts of *Subsection 1.1.1* and *Subsection 1.1.2* are borrowed from [3] and [S4], where the description of optical networks follows closely references [4-9].

1.1.1 Optical Transport Networks

A common state-of-the-art optical transport network with its different hierarchical coverage areas is shown in Fig. 1.2. The network planes represent different levels of aggregated traffic and inter-node distances.

A wide-area network (WAN, global backbone, long-haul network) is a network, that spans over large geographical areas and that carries aggregated global network traffic. In optical communications it represents any terrestrial or undersea system. The reach in such networks is between hundreds to thousands of kilometers. The topology in WAN is typically a mesh. In a mesh topology, every node has a circuit connecting it to neighboring nodes. Mesh topologies are very expensive to implement but yield the greatest amount of redundancy. The bit rate between two WAN nodes is about a few Tbit/s. One standard single-mode fiber (SMF: transport medium in WAN) carries several wavelength channels (e.g., up to 80 wavelength channels) at 10 Gbit/s, 40 Gbit/s or even 100 Gbit/s. The channels are spaced (separation of center carrier frequencies) by 25 GHz…250 GHz, and form a so-called dense wavelength-division-multiplexing (DWDM).

Core switches, which perform switching and routing functions, are typically implemented using optical cross-connects (OXC). OXC interconnect an arbitrary number of input ports with an arbitrary number of output ports. N_f fibers are input to an OXC and then the signal in each fiber is demultiplexed onto M_f separate wavelength channels. Afterwards the M_f channels are guided into an electrical or optical switch fabric and directed onto the desired output port. The switch fabric in the OXC most likely is optical. If it is an electrical switch fabric, no further electrical multiplexing or signal processing is done. The fabric only performs switching of an input signal into an output port, the same way an optical switch fabric would do.

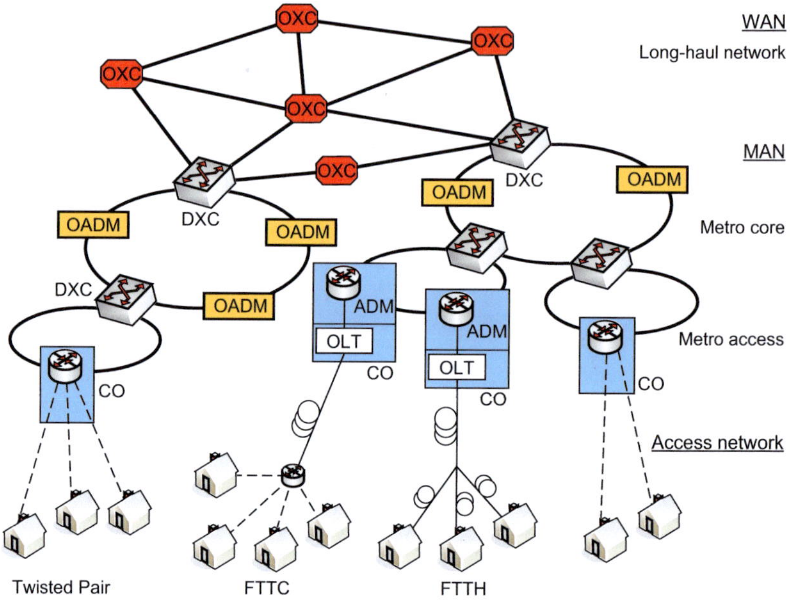

Fig. 1.2: Schematic of a common state-of-the-art optical transport network with its different coverage areas from long-haul to access networks. A wide area network (WAN) has typically a mesh topology connecting terrestrial or undersea systems. The reach of such networks is in the order of hundreds to thousands of kilometers. The metropolitan area networks (MAN, metro) comprise metro-core rings (typical circumference 100…200 km) and metro-access rings (typical circumference 30…60km). The last-mile network, close to the user premises, is called access network. In WAN several wavelength channels are transmitted in a single fiber, and switching of the data signals is done by means of optical cross-connects (OXC). In the metro networks mostly slower and energy-inefficient digital cross-connects (DXC) are used to aggregate and switch the traffic from the WAN to the MAN and also to switch data between the metro-core and the metro-access rings. Optical add-drop multiplexer (OADM) and electrical add-drop multiplexer (ADM) allow a number of channels to bypass a node while only a few channels will be dropped. In the access network, the data signals are sent from the central offices (CO) to the customers and vice versa. Here, fiber (fiber-to-the-*x*, FTT*x*, *x* stands for C: Curb, B: Building or H: Home) and also copper cables (twisted pair) are used to transmit the signals. Modified from [3].

A metropolitan-area network (MAN, metro) is topologically situated between the WAN and the access network plane. It serves the service providers to transport traffic within their own network and serves as an access network for the WAN. The most common metro network topology is a bidirectional ring due to the inherent failure protection. MAN or metro networks are subdivided into:

Metro core: A ring network with a typical circumference of 100-200 km which carries data on several wavelength channels (e.g., 40 wavelength channels) with a bit rate of 2.5 Gbit/s or 10 Gbit/s in a DWDM spacing. The physical layer transport medium is the SMF. Edge routers are used to perform traffic aggregation, switching and routing between the WAN and the metro-core ring. Aggregation and switching are also known as grooming.

The edge routers can be realized using digital cross-connect switches (DXC). In the DXC, optical signals are converted into the electronic domain, electrically switched, and then converted back into an optical signal. Typical DXC has N_f input fibers. Each of the N_f input fibers carries M_f optical wavelength channels, which are first demultiplexed onto M_f separate channels. Since electronic switch fabrics process data at 2.5 Gbit/s [5] and up to 40 Gbit/s [10], further electrical demultiplexing could be required after the optical receiver. The advantage of grooming is that channels can be redistributed with a fine granularity (channel capacity / baseband data rate). The disadvantage is that, e.g., a single 40 Gbit/s channel can require the handling and electronic switching of up to 16 individual lower bit rate channels. DXC offers re-timing, re-shaping and re-amplification (3R) of the data signal. This is known as regeneration. Add-drop multiplexers (ADM) allow a number of channels to bypass a node while only a few channels will be dropped. Of particular importance are reconfigurable optical add-drop multiplexers (ROADM), where optical signals that bypass the switch stay in the optical domain and a certain number of channels can be dropped or added.

Metro access: A ring network with a typical circumference of 30…60 km which carries a few wavelength channels (e.g., 1…16 wavelength channels) with a bit rate of 0.622 Gbit/s…1 Gbit/s in DWDM or coarse WDM (CWDM, 20 nm carrier wavelength spacing). The transport medium is the SMF. DXC are used to perform traffic aggregation and switching between the metro-core ring and the metro-access ring. At the edge to the access network, an electronic ADM aggregates low bit rate traffic coming from the, e.g., local area networks (LAN), enterprise clients, or residential customers.

The access network is the part of the public switched network that connects a central office (CO) or the point of presence with the customer premise equipment (CPE) of subscribers. For instance, the access network includes the subscriber loops and the fiber-to-the-*x* (FTT*x*) access networks (*x* stands for C: Curb, B: Building or H: Home). The traditional transport medium in access networks is the twisted pair (TP) copper cable, known as the subscriber loop. The subscriber loop might be an analog subscriber line or a digital subscriber line (DSL). In recent years, the fiber penetrates towards the CPE with the FFTx approach. Here, in an optical line terminal (OLT), the electrical signals are converted into the optical domain and then send to the CPE. Current maximum downstream bit rates at the customer's side are up to 100 Mbit/s for copper cables and up to 1 Gbit/s for the FTTH approach.

1.1.2 Wired Access Network Implementations

This *Subsection* follows closely the discussion in the books and papers [6, 7, 9]. Today, advanced DSL techniques are deployed using twisted pair and coaxial cable modem techniques. Wired access networks based on copper as the transmission medium are currently reaching their capacity limits. A higher capacity in copper-based networks can only be achieved at the expense of shorter reach or an increased complexity. Thus, a significant increase in reach and capacity is only achievable with optical fibers. Fiber-based networks offer an inherently huge bandwidth, and low losses. Thus, the trend is to replace the old copper cables by optical fibers in the feeder line from the CO to the customer premises. Currently most of the access networks employ both fiber and copper cables. Fibers connect the CO with a cabinet put at the street curb. Here, the fiber is handed over to the copper-cable network. These so-called FTTC solutions will be replaced in the near future by purely fiber-based optical access networks. This approach is the so-called FTTH.

Fiber-Cable Access Networks (FTTC, Fiber DSL Networks)

In the FTTC approach, a feeder fiber connects the CO, i.e., the OLT, with a cabinet housing at the curb site, see Fig. 1.3(a). There, a number of modems support point-to-point connections to a customer premise. Twisted copper pairs can still enable these connections from the modems to the user locations for asymmetric digital subscriber line (ADSL) techniques with a typical length of about a few kilometers (typical bit rate per user about 6 Mbit/s). FTTC can also be used for very high-capacity digital subscriber line (VDSL) techniques with a reach of several hundred meters and a typical bit rate per user of about 50 Mbit/s. These copper lines suffer from bandwidth limitations and from the cross-talk between the copper pairs as multiple lines are combined in a single cable [9]. These effects are counteracted by the latest innovations in the VDSL technology utilizing bonding and vectoring options (offering up to 100 Mbit/s per subscriber).

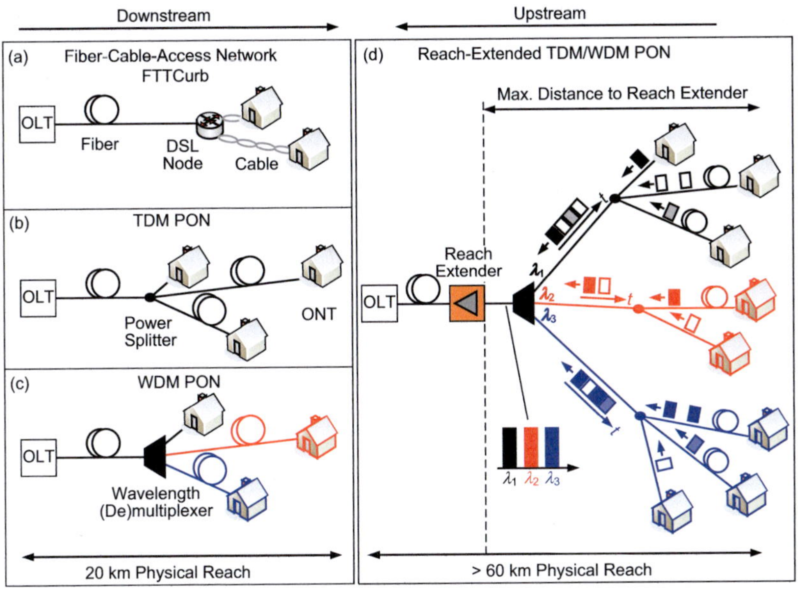

Fig. 1.3: Fiber-based access networks with their different physical reach. (a) FTTCurb scenario, (b) time-division multiplexing (TDM) passive optical network (PON), (c) wavelength-division-multiplexing (WDM) PON and (d) reach-extended hybrid WDM/TDM PON. Modified from [3].

Fiber-to-the-Home (FTTH)

High capacity connections exceeding 100 Mbit/s per home can be realized with a network based on fibers all the way to the home. Several topologies can be implemented which are discussed in the following.

Point-to-Point FTTH: In the point-to-point (P2P) FTTH approach an individual fiber runs all the way from the CO to the user locations. This way there is no competition for capacity among different users. If a single user needs a bandwidth increase, only one specific user needs an upgrade. Moreover, simple Ethernet P2P transceivers can be used, which are becoming very cheap. However, there is a lot of such line-terminating equipment needed, not only at the homes, but also in the CO, which requires more effort for housing, powering and cooling. Moreover, also many fibers need to be installed in the field.

Point-to-Multipoint FTTH: Savings on the feeder part of the fiber infrastructure can be obtained by deploying a point-to-multipoint (P2MP) topology, thus sharing among the users the major part of the infrastructure, namely the feeder line. A fully passive point-to-multipoint topology is the so-called passive optical network (PON) shown in Fig. 1.3(b). It offers the advantage of sharing the feeder fiber and the line-terminating equipment in the CO. By this considerable cost savings are made on the installation and maintenance. As there is no active equipment in the field, there is no need for powering, for equipment which has to be able to withstand large temperature variations, and for expensive street cabinets. Only an optical power splitter is needed in order to distribute the light signal from the CO to the homes. The $1:x$ power splitters equally distribute the power from one input port to x output ports. Currently, power splitters with a split ratio of 1:16 (split losses of 12 dB + 1 dB excess loss) are in use. In the future, 1:512 (split losses of 27 dB + up to 5 dB excess loss) or even higher split ratios are expected.

Access Techniques in PONs

In PONs two access techniques are currently used for sending and receiving data signals.

Time-division-multiplexing (TDM) PON: In a TDM PON the capacity of the feeder fiber can be shared by multiple homes. The downstream direction is traditionally operated at a wavelength around 1490 nm. The OLT at the CO sends the data packets over the PON feeder fiber and thus broadcasts the data to all homes, see Fig. 1.3(b). Upon receipt by the optical network termination (ONT) at each home, the address field of the packets is inspected. If matching with the address of the home, the packet is accepted and delivered to the user. An appropriate medium access control (MAC) protocol is needed to give each customer a fair share of the feeder's capacity.

The upstream direction is typically operated at a wavelength around 1310 nm. A timeslot is assigned to each home in which it is allowed to send. By carefully synchronizing the timeslots of all homes, a collision of the packets at the optical power combiner is avoided. In order to assign the right amount of capacity to each home, the ONT may first send a request packet to the local exchange. The local exchange may grant one (or multiple) timeslots to the ONT. By this, the upstream capacity per ONT can be adapted to its actual traffic load.

The synchronization needed among the upstream timeslots requires that the propagation time from each ONT up to the OLT is precisely known. These times are generally different, as each ONT is at a different distance from the OLT. The distance are determined by sending ranging grants from the OLT, and measuring the roundtrip time upon receipt of these ranging grants returned by each ONT. Hold-off times are inserted at each ONT to equalize the length differences. Every ONT is put at virtually the same length from the OLT, and can then insert its packets in the right time-slot.

All ONTs are usually equipped with the same upstream lasers. However, different optical path lengths between the different ONTs and the OLT and/or different split ratios in the network result in different average optical power levels of the different ONT data packets arriving at the OLT receiver. This leads to a burst signal in the upstream direction, see Fig. 1.4. The ratio between the average optical powers of the data packets is termed burst ratio (BR). The burst-mode receiver in the OLT requires a fast clock extraction and decision level setting. The OLT and the ONT equipment needed for time-division-multiplexing access (TDMA) can be implemented in digital electronics and at relatively low costs.

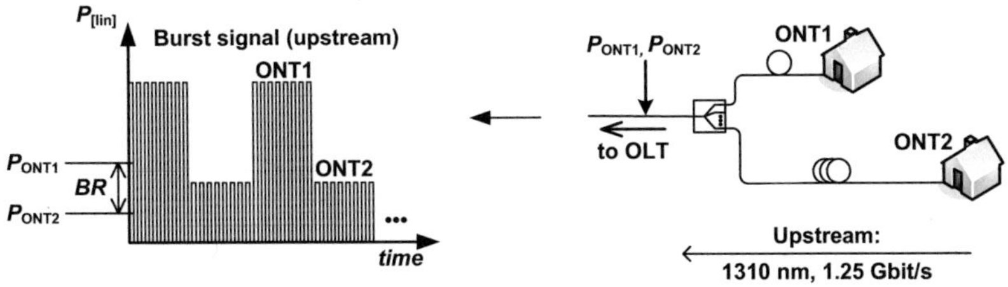

Fig. 1.4: Schematic of the burst signal in the upstream direction of a TDM-PON. Different distances of ONT1 and ONT2 to the power combiner result in data packets with different average optical power levels P_{ONT1} and P_{ONT2} after the combiner and finally at the OLT. The burst ratio (BR) describes the ratio between the average power levels of the data packets. The envelope of the burst signal on the left hand comes from the time-interleaving of the data packets of different ONTs. The OOK data stream is indicated within the packets.

So far, TDMA is the most popular MAC protocol for PONs. It has been implemented in the standardized schemes for broadband PON (BPON), Ethernet PON (EPON), and gigabit PON (GPON). In GPON downstream speeds of 2.5 Gbit/s (about 100 Mbit/s per subscriber) with on-off keying (OOK) modulation are typically used [9]. The maximum reach of a TDMA GPONs is 20 km with a total loss budget of 28 dB (Class B+) [11]. In recent years, next-generation PON1 (NG-PON1) access systems were developed which are based on a TDM-PON approach with a downstream bit rate of 10 Gbit/s. NG-PON1 networks offer up to 1 Gbit/s downstream speed per subscriber.

Wavelength-division-multiplexing (WDM) PON: In WDM-PONs, each subscriber uses its own wavelength channel. In the splitter point of the PON, a wavelength multiplexer/demultiplexer is used to combine/separate these wavelength channels into/from the feeder fiber, see Fig. 1.3(c). No synchronization between the channels is needed, thus they may transport signals with very different formats. In this way, a WDM-PON represents a virtual P2P connection between the CO and each ONT on a shared point-to-multipoint physical infrastructure. Moreover, as the WDM multiplexer performs wavelength routing. The losses in the network splitter point are much lower than in the power splitting PON. For example, in a 1:32 power splitter the losses are about 15 dB, whereas in a 32 channel WDM device the insertion loss can be less than 3 dB per port. Hence the link power budget is considerably better, and thus the reach of a WDM PON can be larger.

As each channel is operating at a specific wavelength, the line-terminating equipment in the OLT at the CO and in the ONT at the user homes needs a wavelength-specific transmitter. This implies that the network operator needs to keep an expensive stock of wavelength-specific modules. A more convenient solution is to deploy universal colorless transceiver modules. Such modules can be realized in various ways. For example, a reflective modulator or reflective semiconductor optical amplifier (RSOA) can be used in the transmitter module at the ONT to generate the upstream data. Upstream data speeds of 1.25 Gbit/s and beyond have been obtained [6, 7, 9].

Next-Generation Reach-Extended PON: A current trend is the convergence of TDM and WDM access schemes into a so called hybrid WDM/TDM PON (e.g., NG-PON2 approach using 4 WDM channels with 10 Gbit/s each in the downstream direction). This enables a large-scale FTTH network. By deploying multiple wavelength channels a large capacity is realized, and by using TDM within each wavelength channel a large amount of users can be achieved. Such hybrid WDM/TDM schemes are currently investigated for long-reach PON systems featuring a reach of more than 100 km with more than 1000 connected homes, see Fig. 1.3(d). The number of COs can be reduced, saving operational cost.

The longer reach with a higher split ratio causes a loss increase. Thus, reach-extender (RE) amplifiers are required in the downstream and in the upstream paths. Especially in the upstream path, the RE box needs to cope with fluctuating input power levels due to the different distances to the customer locations. The maximum input power variation is determined by the power difference between the ONT with the maximum distance to the RE and the ONT with the minimum distance to the RE. The input power variation in today's TDM PON is about 10…15 dB [11]. However, this power variation will increase significantly with an increase in network complexity. It is expected that input power dynamics in the range of 40 dB are needed in next-generation access networks.

In next-generation long-reach PONs, the RE will also have to cope with different high-speed wavelength channels simultaneously, and with advanced optical modulation formats. Overall requirements for a RE box are that it needs to withstand summer heat as well as winter cold, when it is deployed in the street cabinet. Capital expenditures (CAPEX) and operational expenditures (OPEX) should be as low as possible to enable quick revenue.

Orthogonal frequency division multiple access (OFDMA)-PON: In OFDMA-PON systems [12-14] orthogonal electrical subcarriers of a so called orthogonal frequency division multiplexed (OFDM) signal are used to transmit data from the CO to the ONT and vice versa. The CO can dynamically assign different subcarriers of the OFDM signal to different ONTs and services in the network. In OFDMA systems the orthogonal subcarriers are processed with low-cost digital signal processing (DSP) which is used to implement the inverse fast Fourier transform and the fast Fourier transform for multiplexing/demultiplexing of the orthogonal subcarriers of the OFDM signal. Several hundreds of electrical OFDM subcarriers may be very densely packed (subcarrier spacing of only some MHz), leading to a very fine granularity in such systems. On each subcarrier complex modulation schemes can be used to encode data.

Reach Extender Technologies

A brief overview of possible reach-extender technologies is given at this point. In this thesis, the focus lies on SOAs which will be introduced in more detail in *Section 2.4*.

Semiconductor Optical Amplifier (SOA): An SOA is a semiconductor waveguide with a gain medium. In an SOA the gain is obtained by injecting carriers from a current source into the active region. The SOA technology can offer a large gain bandwidth about 50 nm…100 nm and almost polarization independent amplification. SOA are available in the spectral range of the downstream (around 1500 nm) and of the upstream (around 1300 nm), respectively. If the devices are operated in gain saturation, bit-pattern effects can occur for high-speed data signals, and cross-talk in multi-wavelength operation. On the contrary, if the devices are operated with low input power signals, the optical signal-to-noise ratio (OSNR) limits the performance. Here, a large dynamic range is desirable to overcome the limitations. SOA can be deployed in a PON within the CO to boost/pre-amplify the data signal which is sent to/from the customers, as an in-line amplifier in the field or within the ONTs.

Raman-Amplifier: If two optical fields are launched into a fiber with a difference frequency of 13 THz, then this matches the optical phonon resonance frequency and a beating of the fields can lead to strong coherent optical phonon oscillations. This process leads to stimulated scattering of the strong light signal into the weak one. This stimulated Raman scattering can be exploited to amplify an optical signal. In practice, a strong pump laser (located at the CO) at a wavelength of 1450 nm is added to the optical signal which is typically at 1550 nm. As both signals have a spacing of 100 nm or 13 THz in terms of frequencies, energy is transferred. The distributed amplification takes place along a distance of 20 km. With this technology bit-pattern effects can be avoided and the occurrence of fiber nonlinearities is significantly reduced due to the distributed amplification process. Furthermore, a low noise figure is possible. However, a low bandwidth of about 20 nm (using a single laser source) and the need for a strong pump laser increases the costs compared to the SOA technology. At a wavelength of 1.3 µm this technology is not as mature as the SOA technology.

x-Doped Fiber Amplifier (xDFA): Fiber amplifiers are usually based on rare-earth doped fibers (Erbium (EDFA) for C-band operation or Praseodymium (PDFA) for O-band operation). In contrast to electrically pumped SOA, fiber amplifiers do not only need a doped fiber but also an optical pump laser. Their gain bandwidth is limited to something between 30 nm...70 nm. Especially in the upstream path of access networks, where the bursty traffic is present (bursts on millisecond and microsecond time scales), fiber amplifiers induce strong signal degradations due to their fluorescent lifetimes (10 ms in an EDFA). Burst capable EDFAs are available, but they need additional equipment [15]. An advantage of fiber amplifiers is their large small-signal gain of up to 40 dB and low noise figures down to 4 dB or less. They are polarization independent and widely used in multi-wavelength applications due to their low channel cross-talk. Patterning effects at high speed applications (10 Gbit/s and higher bit rates) are avoided due to the long fluorescent lifetimes. *x*DFA are available at moderate cost. At a wavelength of 1.3 µm this technology is no mass product.

Optical-Electrical-Optical (OEO) Units: In active access networks the reach can be extended with OEO units. The advantage of an OEO conversion is that data signals are format-transparently regenerated and errors are corrected. However, an OEO unit is expensive, complex, has a large footprint, and large power consumption. If the bandwidth of an access network requires an upgrade, the OEO units need to be replaced, which is a time consuming and traffic-disruptive procedure.

Coherent Transceiver: The use of phase-modulated signals in conjunction with coherent reception at the OLT and ONT sites can significantly increase the access network reach. The overall power budget is increased due to the fact that lower receiver input power levels are tolerable compared to direct-detection schemes [16].

1.1.3 Mobile Backhaul Network

As pointed out at the beginning of this *Chapter*, the mobile data traffic will dramatically increase within the next years. The end-user demands high data rate and high mobility at reasonable prices. To cope with this demand, the capacity of legacy wireless networks needs to be continuously upgraded and future-proof wireless access technologies have to be introduced. Over the last decades, different generations of cellular telecommunication networks have evolved and the capacity per radio cell has increased dramatically. The numbers given in the following were obtained from [17]:

The second generation (2G) mobile networks are based on the GSM (global system for mobile communications) standard and use time division multiple access (TDMA) or code division multiple access (CDMA). The GSM carrier frequencies are 900 MHz and 1800 MHz. In 2G networks only voice and text services (short message service, SMS) are supported. An extension of GSM is EDGE (enhanced data rates for GSM) also denoted as 2.75G networks. EDGE offers additional services such as multimedia messaging service (MMS) and internet access. The peak network speed is 237 kbit/s in both the uplink and downlink direction. The typical average user rate is 70…135 kbit/s in both the uplink and the downlink.

A multitude of additional services including mobile TV and fast internet access is offered by third generation (3G) mobile networks. The main standards of 3G networks are UMTS (universal mobile telecommunications system), HSPA (high speed packet access), and CDMA2000 (predominantly America, Asia, and Africa). UMTS and HSPA employ CDMA and use the 2100 MHz carrier band (Europe). The peak network speed is 384 kbit/s (typical user rate of 200…300 kbit/s) for UMTS in the uplink and in the downlink. With HSPA a peak network speed of 14 Mbit/s can be achieved in the downlink. However, typical user rates are in the order of 2 Mbit/s. An improved version of HSPA, HSPA+, offers theoretical peak downlink speeds of up to 42 Mbit/s and expected user rates > 5 Mbit/s.

The fourth generation (4G) mobile communications standard is called LTE (long term evolution) and LTE-Advanced [18], respectively. LTE uses OFDMA for the downlink and SC-FDMA (single-carrier frequency-division multiple access) for the uplink. SC-FDMA is a similar access scheme to OFDMA, however, a more power efficient one. LTE uses different frequency bands in different countries (e.g. 800, 900, 1800, and 2600 MHz in Europe). With LTE (LTE-Advanced) maximum downlink and uplink speeds in the order of 300 Mbit/s (1 Gbit/s) and 75 Mbit/s (1 Gbit/s) are possible.

Another 4G wireless technology is WiMAX (Worldwide Interoperability for Microwave Access). WiMAX operates at frequencies in the 2 to 66 GHz range, uses OFDMA and provides downlink speeds in the order of several tens and hundreds of Mbit/s. WiMAX is considered as an alternative for GSM, UMTS, and LTE mobile technology but also for fixed copper wire connections.

The increasing mobile data traffic at the cell site, inevitably results in an increase of data traffic in the connected transport networks. In this context the mobile backhaul network usually is a major bottleneck. The mobile backhaul connects the base stations of the radio access network (RAN) with their associated controller or gateway sites. The controller or gateway sites represent the border between the RAN and the mobile core network or the internet.

Fig. 1.5 shows a simplified schematic of mobile backhauling in 2G, 3G, and 4G networks [19, 20]. Elements with similar tasks in the various network types have different nomenclature. In the 2G network, the base station is denoted as the base transceiver station (BTS), whereas in 3G and 4G networks these elements are called Node B and evolved (e)Node B, respectively. For the mobile user, the base station represents the access point to the rest of the network. The traffic of several base stations is aggregated and then further directed to the associated controller and gateways which are the base station controller (BSC) in 2G networks, the radio network controller (RNC) in 3G networks, and the gateway (GW) in 4G networks, respectively.

The common physical access technologies in the RAN are copper wire, microwave, or fiber. The choice of technology depends on several factors such as cost, required capacity, reach, and site availability. Fiber access offers Gbit/s to Tbit/s capacity and is best suited for backhaul links over longer distances. However, fiber infrastructure is not available everywhere and setting up new a fiber infrastructure is expensive.

Historically, the mobile networks rely on TDM transport services for interconnection between the base stations and the controllers. The physical backhaul technology mainly used in legacy 2G and 3G networks are copper wire (E1/T1, dashed lines in Fig. 1.5) and microwave (yellow flash symbol in Fig. 1.5). Both solutions offer a small backhaul capacity compared to fiber. The conventional microwave backhaul operates in frequency bands between 6 GHz and 38 GHz and provides up to 500 Mbit/s single carrier backhaul capacity. The capacity can be further enhanced using multi carrier operation and/or multiple-input multiple-output (MIMO) techniques. Meanwhile the mm-wave frequencies in the 71 to 76 GHz and 81 to 86 GHz band (E-band) are used for wireless backhauling solutions with Gbit/s capacity with simple modulation formats on a single carrier.

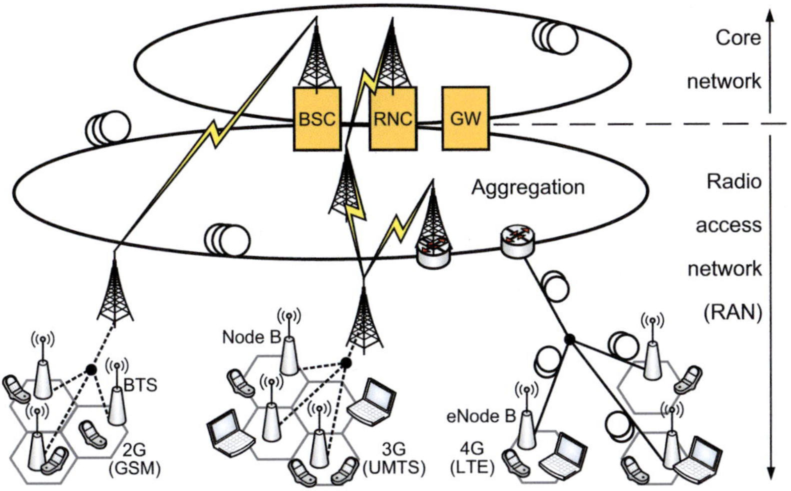

Fig. 1.5: Mobile backhauling in 2G, 3G, and 4G networks. The base stations (base transceiver station (BTS) in 2G mobile networks, Node B and eNode B in 3G and 4G mobile networks) are the access points in the radio access network (RAN). The mobile backhaul connects the base stations of the RAN with their associated controller or gateway sites, i.e., with the base station controller (BSC) in 2G networks, the radio network controller (RNC) in 3G networks, and the gateway (GW) in 4G networks. The controller or gateway sites themselves represent the border between the RAN and the mobile core network or the internet. The physical backhaul technology in legacy 2G and 3G networks are mainly copper wire (dashed lines) and microwave (yellow flash symbol). Optical fiber (solid line with coil symbol) is the only option as physical backhaul technology in 4G networks due to the large amount of data. The aggregation part of the mobile backhaul network is based on an optical fiber infrastructure (Ethernet ring) or high capacity point-to-point wireless links.

In 4G networks the LTE backhaul capacity per base station is estimated to exceed 90 Mbit/s [19]. The required backhaul capacity easily sums up to several Gbit/s when multiple LTE base stations are aggregated. In practice, this virtually leaves fiber (solid line with coil symbol in Fig. 1.5) as the only suitable physical access technology for LTE. In 2012, T-Mobile stated that 95 % of their 4G backhaul is fiber [21].

The aggregation part of the mobile backhaul network is commonly based on optical fiber infrastructure (Ethernet ring) or high capacity P2P wireless links. WiMAX is also considered as an option for mobile backhauling.

1.2 Network Convergence

Current optical and mobile access networks were reviewed in the previous *Section*. In this *Section*, the requirements of future optical and mobile access networks are discussed in the context of network convergence.

Network convergence results from the convergence of services and contents that are demanded by the end-user. Today, the end-user would like to have ubiquitous and simultaneous access to voice, data, video, and interactive multimedia services. From a network perspective this requires that (a) the network supports all of these services, (b) the network provides enough capacity to provide these services and (c) this high-capacity network infrastructure reaches up to the end-user.

For the first requirement, an IP based end-to-end network is needed, which enables the end-user to access all kinds of services over one or more networks. The second and third requirements, a high-capacity network with large area coverage, affect several network aspects at the same time. Primarily, this implies the convergence of optical and wireless networks, but also the convergence of the metro and access network infrastructures.

1.2.1 Converged Optical-Wireless Network

LTE and LTE-Advanced are broadband wireless access solutions that offer mobility to the end-user. However, the amount of mobile traffic generated in LTE networks must also be supported by the mobile backhaul. In case of LTE-Advanced, a wireless backhaul solution based on microwave or copper hardly can provide sufficient capacity. In contrast, fiber-based access techniques as discussed in *Section 1.1.2* virtually offer unlimited bandwidth.

Optical-wireless convergence [22] aims at sharing the same infrastructure for fixed and wireless services. A so called radio-over-fiber (RoF) system [23-25] combines the advantages of huge bandwidth and low transmission losses in fiber-based access networks with the ubiquity and mobility of wireless access networks. In a RoF system, a radio signal is converted to an optical signal (electrical-to-optical conversion, E/O) for transmission over an optical fiber network and finally converted back to the radio signal (optical-to-electrical conversion, O/E). The easiest and cheapest RoF technology employs intensity modulation (IM) and direct detection (IM-DD) [26]. A RoF link uses small-signal modulation to encode the radio signal on the optical carrier. Thus small-signal models of the different RoF link components are used to describe the link. This is in contrast to traditional optical communication links using digital on-off-keying modulation where the light is either on or off and where the link is described with a large-signal model.

The increasing number of mobile users accessing broadband services leads to a shrinkage of the radio cell size to micro-cells (cell radius 2km), pico-cells (200 m) or even femto-cells (10 m). In small radio cells, mm-wave carrier frequencies (e.g., in the 60 GHz-band) which may carry Gbit/s data can be used. The mm-wave signals can be transmitted and distributed over several kilometers distance using RoF technology with low fiber attenuation. In coax cables or air, the distance would be limited to only a few meters due to the high attenuation.

Optical-wireless (or fixed-mobile) convergence also enables new techniques like coordinated multi point (CoMP) [27-29] in LTE systems, where cooperating base stations significantly increase the wireless transmission capacity. The optical fiber provides the means for interconnecting different base stations in a CoMP system.

From the discussion above, it becomes clear that optical fiber is the transmission medium of choice in future access networks and for mobile backhauling solutions. The fiber infrastructure needs to come as close as possible to the end-user, i.e., FTTH connections for the fixed-wired access and fiber-to-the-base station for the mobile access in a converged optical-wireless network.

The reach in a converged optical-wireless network can be extended by SOAs. In this thesis, the use of SOAs with RoF signals is investigated. Specific SOA requirements for such systems are discussed in *Section 1.3*.

If an end-to-end fiber connection is not available, a high-capacity mm-wave wireless link can be an adequate alternative. One part of this thesis investigates the integration of a high-capacity mm-wave wireless link into an optical fiber infrastructure. Specific aspects and requirements for such a scenario are elaborated in *Section 1.4*.

1.2.2 Converged Metro-Access Network

To account for the various requirements of network convergence, also novel optical network topologies have to be considered. In this context, a *converged metro-access network* [30, 31] suits better the needs than common networks which are traditionally structured into a metro-core network, a metro-access network, and an access network (see Fig. 1.2). In a converged metro-access network infrastructure, as schematically depicted in Fig. 1.6, large parts of the metro network and the access network have merged into a common infrastructure, the metro-access ring. The fiber ring has a circumference of several tens of kilometers. The main CO is located at the edge between metro core and converged metro-access network. In each CO, an OXC performs switching and traffic aggregation. Different metro-access rings may be linked by an OXC. By this means, access networks which are attached to different metro-access rings are virtually connected which each other.

On the metro-access ring, different wavelength bands (20 nm bandwidth, CWDM) in the range of 1.3 μm and 1.5 μm are used. Each wavelength band contains several subchannels. ROADMs along the metro-access ring represent the access point to the ring. The ROADMs provide reconfigurable and/or fixed add/drop ports where either active system equipment or other infrastructure segments can be connected, e.g., a WDM/TDM PON for FTTH, a fiber link for mobile backhauling (RoF), or a direct fiber connection to enterprise customers, data centers, and large institutions such universities or research centers.

The converged metro-access network approach offers high capacity (fiber-based) in a large coverage area (long reach). The flexibility and versatility of the network is further enhanced, if the infrastructure extends transparently over larger areas (large cities or regions). Thereby even more services are attached to the same network. In principle, the infrastructure can be operated and controlled by a single provider. However, it may also be opened for multiple providers that coexist and share the same infrastructure [31].

SOAs are used in each of these infrastructure segments to extend the overall reach of the converged metro-access network, e.g., in a long reach (100 km) TDM/WDM approach with a high split ratio of 1:512. SOAs are also included in each ROADM on the metro-access ring. Separate SOAs are used for each wavelength band to compensate for the ROADM losses and to increase the overall loss budget. In this thesis, the use of SOAs in a converged metro-access network scenario is investigated. The SOA requirements are elaborated in *Section 1.3.*

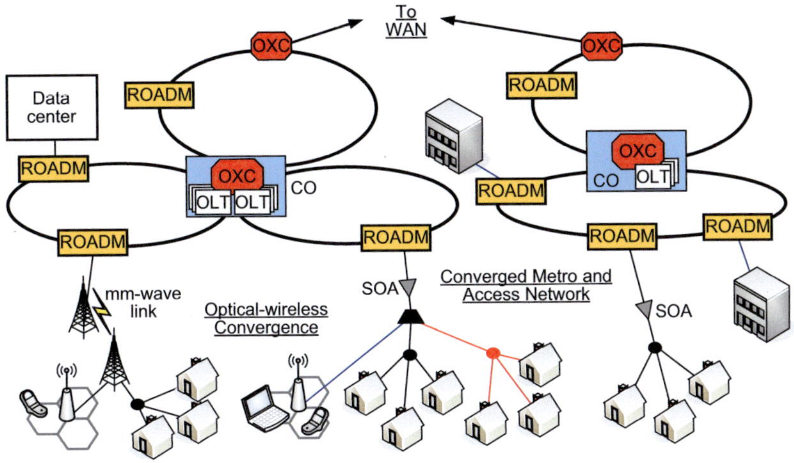

Fig. 1.6: Schematic of a converged metro-access network infrastructure. Large parts of the metro network and the access network have merged into a common infrastructure, the metro-access ring with a circumference of several tens of kilometers. Different metro-access rings may be linked by an OXC. ROADMs along the metro-access ring provide reconfigurable and/or fixed add/drop ports where either active system equipment or other infrastructure segments can be connected, e.g., a WDM/TDM PON for FTTH, a fiber link for mobile backhauling (RoF), or a direct fiber connection to enterprise customers, data centers, and large institutions. SOAs are used in each of these infrastructure segments and in the ROADMs to extend the overall reach of the converged metro-access network. Modified from [3].

1.3 SOAs in Converged Access Networks

The trend towards network convergence and optical-wireless convergence was discussed in the last *Section* and some advantages of SOAs were pointed out earlier in this *Chapter*. In this thesis, the use of SOAs as loss compensating elements in converged access networks is studied. The central question in this context is, whether the SOA distorts the transmitted signal, and if yes, how strong is the distortion and which limitations result from this for the use of SOAs in certain access network scenarios. Thus, the optimum operating conditions for SOA must be identified for different kind of modulation formats and multiplexing techniques.

1.3.1 Cascaded SOAs in Metro-Access Ring Networks

One scenario investigated in this thesis is an optical metro-access ring network with SOAs as loss compensating elements in ROADMs. Such a scenario is shown in Fig. 1.7 (left).

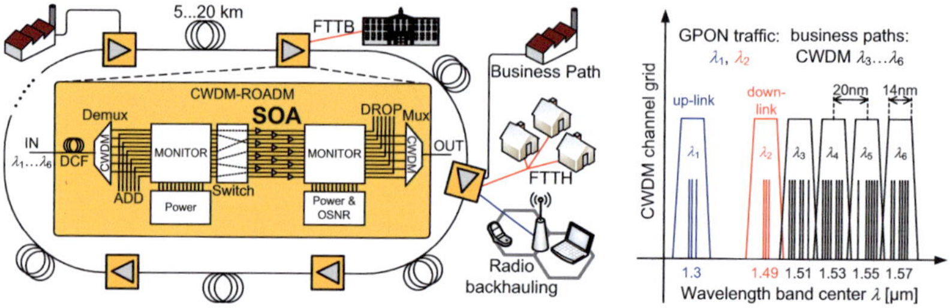

Fig. 1.7: Converged metro-access ring network scenario. The loss compensating SOAs are located in the ROADMs which provide data switching every 5 km…20 km. This network supports infrastructures such as fiber-to-the-x (FTTx with x = B, H for building, home) or radio backhauling using 14 nm wide bands (20 nm separation) centered around $\lambda_1 \ldots \lambda_6$. Two wavelength bands are reserved for upstream and downstream traffic in a GPON. Four wavelength bands are for high capacity business paths or other infrastructure segments. Each band contains several subchannels and is amplified by a dedicated SOA. The signals may be transmitted over a chain of ROADMs, thus a *cascade of SOAs* results. Each subchannel in a wavelength band can carry data employing different multiplexing techniques, modulation formats and channel powers. Modified from [C11].

The network provides sufficient bandwidth for different kinds of customers and services such as GPON, FTTx, high-capacity "business paths" for enterprise customers, and radio backhauling. The ROADMs (< 20 km distance between two ROADMs) along the ring make the network versatile: A ROADM supports 6 coarse wavelength division multiplexing (CWDM) bands ($\lambda_1 \ldots \lambda_6$) each separated by 20 nm and having a bandwidth of 14 nm for several subchannels, Fig. 1.7 (right). Two wavelength bands are reserved for upstream and downstream traffic in a GPON, and four wavelength bands are for high capacity business paths. Each ROADM comprises a dispersion compensating fiber (DCF) module, a demultiplexer/multiplexer (DEMUX, MUX) and monitoring taps (MONITOR) for power and optical signal-to-noise-ratio (OSNR) monitoring. The switch (SW) offers the flexibility to add and to drop an individual wavelength band, or to put it through to the next ROADM.

In each ROADM, each CWDM band is amplified by a dedicated SOA to compensate for the ROADM losses and the transmission losses along the ring. The data signals may be transmitted over a chain of ROADMs in such a ring network, thus a *cascade of SOAs* results. Since each subchannel in a wavelength band can carry data employing different multiplexing techniques, modulation formats and channel powers, the question is, if the data signals are distorted by the SOA cascade.

Previous works [32-34] already investigated SOA cascades with OOK modulation. Today, advanced modulation formats that are spectrally more efficient than OOK modulation are of interest for future access networks. Advanced modulation formats encode the data in the amplitude and the phase of the optical field [35]. So far, a cascade of SOA employing advanced modulation formats has not been investigated. In *Section 3.1* of this thesis, a cascade of commercially available quantum well (QW) SOAs is studied for single channel and multi-channel signals carrying advanced modulation formats at speeds > 100 Gbit/s. In *Section 3.4.2*, an SOA cascade of quantum dot (QD) SOA is investigated and the results of a re-circulating loop experiment with broadband RoF OFDM signals are presented.

It should be noted that the emphasis of this thesis is not to compare the performance of QW SOAs with that of QD SOAs. The choice of the SOA was mainly dependent on the availability of the respective device(s) at the time of the experiment. However, in the rare cases where there is evidence for a different behavior of both types of SOA under certain operation conditions, it is explicitly pointed out.

1.3.2 SOA Reach-Extender in Optical-Wireless Networks

The second type of scenario that will be investigated in this thesis is the use of an SOA as a reach extender in converged optical-wireless networks as depicted in Fig. 1.8. The figure shows a schematic WDM/TDM PON converged with RoF applications with an SOA as reach extender between the CO and the WDM multiplexer. The data signal could also come/go from/to a ring network (ROADM) instead of the CO.

One question is, if the SOA distorts a RoF signal (small-signal modulation) and to which extent variations of the RF power do affect the signal quality when amplified by an SOA. Different RF power levels could for instance occur at the mobile base station which translates into different signal strengths of the radio signal that is modulated on the optical carrier. The small-signal RF response of an SOA is determined in *Section 3.2.1*. Furthermore, the electrical power dynamic range (EPDR) of a SOA for different kind of RoF signals is investigated in *Section 3.2.3*, *Section 3.3.2*, and *Section 3.4.1*. The EPDR is defined as the power range of the RF signal modulated on the optical carrier (i.e., the range of the optical modulation depth) within which error-free operation is obtained for a given average optical SOA input power.

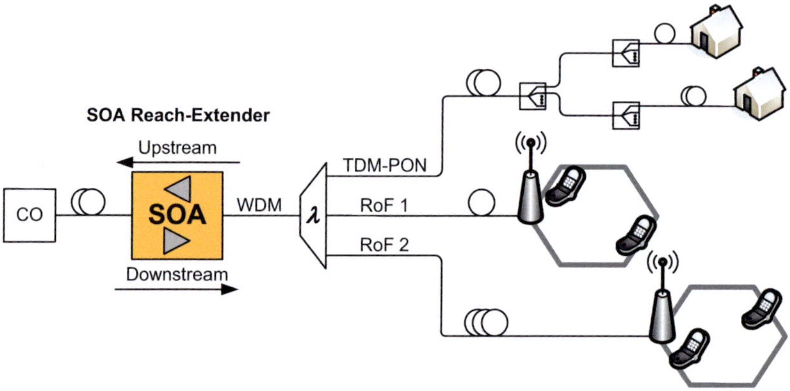

Fig. 1.8: SOA reach-extender in a converged optical-wireless network. A WDM/TDM PON is converged with RoF. The SOA simultaneously amplifies the different types of data signals. The SOA must offer a large optical and electrical power dynamic range to guarantee a functional network.

However, also the optical power dynamic range (OPDR, also known as the input power dynamic range, IPDR), is of great interest, especially in the upstream direction. The OPDR is defined as the optical SOA input power range within which error-free operation is obtained. Different distances of the ONUs of the TDM-GPON to the reach-extending SOA and/or different split ratios in the PON vary the optical input power to the SOA. Moreover, different fiber lengths from different radio base stations to the SOA vary the optical input power of the RoF signals at the SOA input. The SOA OPDR for RoF signals is also studied in *Section 3.2.3* and *Section 3.3.2*.

Another question is, how a RoF signal is influenced by other signals, e.g. a second RoF signal or a bursty TDM-GPON signal, if all signals are amplified simultaneously by the same SOA, see scenario in Fig. 1.8. The use of an SOA as a reach extender in the upstream path and the impact of a rival RoF, a rival WDM, and a rival bursty TDM-PON signal on a RoF subscriber are investigated for a converged optical-wireless network in *Section 3.3*.

1.4 High-Capacity mm-Wave Link in Access Networks

In the course of increasing IP traffic and network convergence, high speed optical fiber connections are continuously pushing towards the network edge and towards the subscribers. However, FTTH is mainly available for subscribers in selected areas, predominantly urban and metropolitan areas. Usually, the "last mile" to the subscriber is still covered with twisted-pair copper wires that significantly limit the data rate and the distance to be bridged. Rural and remote areas are often completely cut-off from broadband internet access, as no costly optical fiber infrastructure is available, and because the data capacity of existing telephone lines is insufficient if longer distances have to be bridged. Providers do not see much interest in an immediate infrastructure investment and prefer densely populated areas.

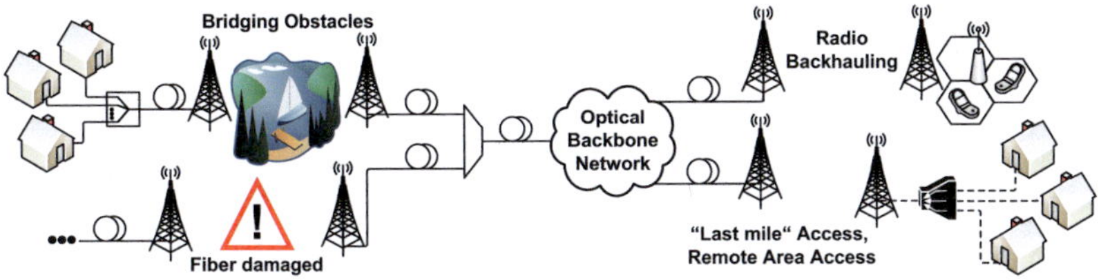

Fig. 1.9: Various applications of a high capacity mm-wave wireless link in a converged optical-wireless network: Connecting fiber-optic data links (e.g. to bridge obstacles such as lakes, valleys, or construction sites), acting as "last mile" access (e.g. in rural and remote areas), an *ad hoc* solution if an existing fiber connection fails, and radio backhauling. In contrast to free space optical links that use visible or infrared light for point-to-point communications, wireless links stay operational also under adverse weather conditions such as heavy rain or fog.

In cases where an end-to-end fiber connection is absent and the deployment of a new fiber link is not economical, a high-capacity wireless connection could help, see Fig. 1.9. A wireless link capable of transmitting data rates comparable to those found in (future) optical access networks (multi-gigabit capacity in the order of 100 Gbit/s) could be an adequate fiber substitute. If such a wireless link is integrated into an optical fiber infrastructure, the wireless link virtually acts as a reach extender of the optical network ("fiber-over-radio"). Such a high-capacity wireless link can also be used for bridging difficult-to-access terrains and natural obstacles, radio backhauling, and as an *ad hoc* solution if an already existing fiber connection fails.

Multi-gigabit wireless transmission asks for multi-GHz bandwidths, which are only available at much larger frequencies than the usual mobile communications uses. Millimeter-wave frequencies (radio frequencies in the range of 30 GHz to 300 GHz) fulfill this need. Fig. 1.10 shows the atmospheric attenuation in the frequency range of 0…500 GHz for different weather conditions [36-38]. Attenuation peaks due to oxygen (O_2) and water (H_2O) absorption can be seen at 60 GHz, 118 GHz, 182 GHz, and 325 GHz. Currently, a lot of research concentrates on wireless communication at mm-waves in the unlicensed 60 GHz band between 57…64 GHz and the W-band between 75…110 GHz.

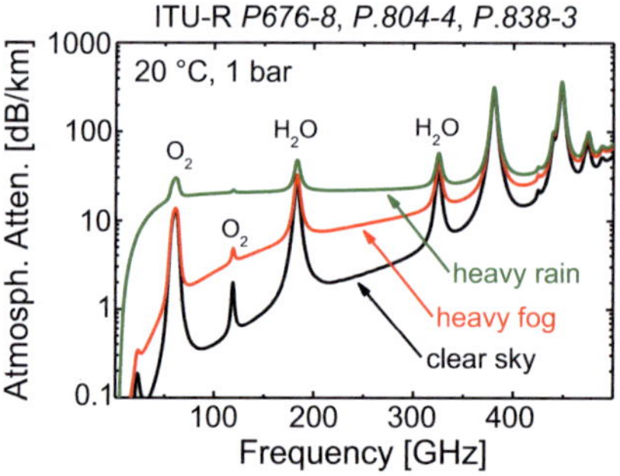

Fig. 1.10: Atmospheric attenuation in the range 0…500 GHz under different weather conditions according to ITU standards [36-38]. Attenuation peaks due to oxygen (O_2) and water (H_2O) absorption can be seen at 60 GHz, 118 GHz, 182 GHz, and 325 GHz. In this work, high capacity wireless links operating in the atmospheric transmission window between 200…300 GHz are investigated.

In this thesis, the integration of a wireless link operating at a carrier frequency of 220 GHz and 240 GHz into an optical fiber system is investigated. As can be seen from Fig. 1.10, there is a very broad atmospheric transmission window between 200…300 GHz, which offers sufficient bandwidth for a 100 Gbit/s wireless link. It can also be seen from Fig. 1.10 that under bad weather conditions (heavy rain) the atmospheric attenuation in this band (about 23 dB/km) is very similar to the atmospheric attenuation in the W-band (75 GHz to 110 GHz). Free-space optical links (FSO) could also provide the required bandwidth. However, in contrast to mm-wave or terahertz transmission, FSO links become virtually inoperative under such adverse weather conditions [39, 40].

Electronic and Photonic mm-Wave Signal Generation

Two different concepts for generating mm-wave signals and integrating the wireless link into a fiber environment are discussed in this thesis:

The first concept is based on *electronic up-conversion*. An optical data signal is transmitted over fiber and then received at a photodetector. After photodetection, the baseband data signal is up-converted to the mm-wave frequency carrier by a broadband electronic mixer implemented with mm-wave monolithic integrated circuits (MMIC) [41], *Appendix A.4*. The mm-wave signal is then radiated over an antenna. At the wireless receiver (Rx), the data signal is down-converted to the baseband employing a MMIC mixer. The baseband signal can then again be encoded onto an optical carrier which is further transmitted over fiber. In *Section 4.1*, a wireless link operating at 220 GHz based on electronic up-conversion is demonstrated to connect two optical links at data rates of up to 20 Gbit/s for OOK or 9 Gbit/s for electrical OFDM. Intensity modulation and direct detection is employed in the two fiber sections, each with a length of up to 20 km.

The second, very elegant method for generating mm-waves is based on *photonic up-conversion*. In this approach, the up-conversion to the mm-wave carrier is done photonically by photomixing (optical heterodyning) a local oscillator (LO) and the modulated optical data signal at a broadband photodetector. The frequency spacing between the LO and the optical data signal defines the generated mm-wave frequency. For generating mm-wave frequencies > 100 GHz, a very broadband uni-travelling-carrier (UTC) photodiode (PD) [42] is employed, *Appendix A.5*. In *Section 4.2* and *Section 4.3*, a record 100 Gbit/s wireless link at 237.5 GHz based on an optical heterodyne IQ transmitter and a state-of-the-art active IQ-MMIC at the Rx is demonstrated. Photonic generation and transmission of single-channel and multi-channel signals is investigated.

State of the Art: Electronic and Photonic Up-Conversion

Fig. 1.11 relates the work on high-capacity mm-wave links presented in this thesis (stars) to the state of the art (circles). The figure shows the achieved wireless data rate as a function of the mm-wave carrier frequency. Some groups use 2×2 multiple-input multiple-output (MIMO) techniques (hollow circles) to double the data rate compared to a single-input single-output (SI-SO) system such as used in this work. To make results comparable, the chart shows always the equivalent SISO data rate. The number next to a symbol gives the wireless transmission distance in meter. If no number is given, no wireless transmission distance was specified or only a back-to-back measurement was performed (transmitter and receiver were directly connected with each other). Black symbols indicate systems employing photonic up-conversion. Red symbols indicate systems with electronic up-conversion.

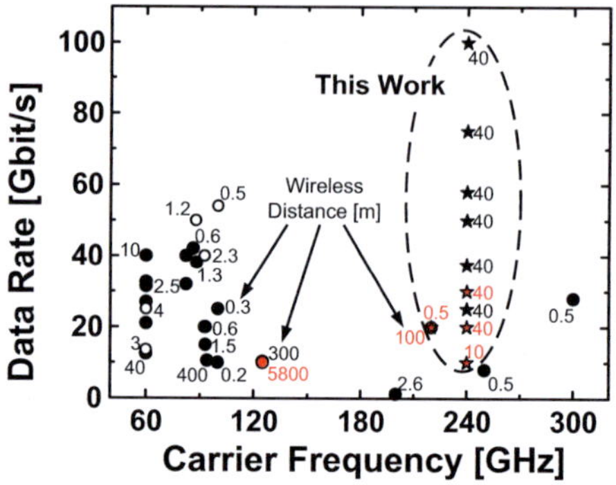

Fig. 1.11: High-capacity mm-wave links presented in this thesis (stars) [J2], [J3], [C1], [C6], [C8], [C10], [C17], [C18] compared to the state of the art (circles) [43-74]. The number next to a symbol gives the wireless transmission distance in meter. No number means that no wireless transmission distance was specified or only a back-to-back measurement was performed. Black symbols indicate systems employing photonic up-conversion. Red symbols indicate systems with electronic up-conversion. Some groups use 2×2 multiple-input multiple-output (MIMO) techniques (hollow circles). In these cases, the chart shows the equivalent single-input single-output (SISO) data rate.

As can be seen from Fig. 1.11, numerous works have been published in the 60 GHz [43-53] band and in the W-band (75-110 GHz) [54-66]. At these frequencies, hardware is commercially available. Photonic up-conversion is the preferred method for generating mm-wave signals. Advanced modulation formats and 2×2 MIMO techniques are employed in the 60 GHz band and the W-band. This led to data rates of 80 Gbit/s [67], 100 Gbit/s [68], and 108 Gbit/s [69] in the W-band. In mm-wave system operating at carrier frequencies beyond 110 GHz, simple OOK modulation and envelope detection with Schottky barrier diodes at the Rx is used [70-72]. Data rates of up to 28 Gbit/s have been shown [73].With electronic up-conversion high RF transmit powers can be achieved, which results in large transmission distances (often also beam-focusing antennas are used). In [74], a 10 Gbit/s OOK signal was electronically up-converted to 125 GHz and transmitted over a wireless distance of 5.8 km.

In this work, a record SISO wireless link at 237.5 GHz based on photonic up-conversion is implemented with data rates of up to 100 Gbit/s. This huge wireless capacity is achieved by accessing the large usable bandwidth around the high mm-wave carrier frequency with state-of-the-art photonic and electronic technology. An ultra-broadband UTC-PD performs photonic up-conversion and after wireless transmission, a broadband active MMIC Rx down-converts the signal to the baseband. Spectrally efficient advanced modulation formats are employed to transmit as many data as possible within the available system bandwidth. A transmission distance of 40 m is achieved in an indoor lab environment (beam-focusing antennas were used). An increase to transmission distances > 1km is feasible with state-of-the-art amplifiers and highly directional antennas.

1.5 Conclusion

The large increase of data traffic in both wired and mobile communication systems leads converged optical-wireless and converged metro-access networks. Due to their large optical bandwidth and low-cost, SOA are promising candidates to be used as loss compensating elements and reach-extender in such converged access networks. In this context, (linear) SOAs offer the following features:

- Large input power dynamic range for the amplification of advanced modulation formats and RoF signals
- Multi-channel signal amplification
- Simultaneous amplification of RoF signals and (bursty) WDM/TDM-PON traffic
- Compatibility with different multiplexing techniques such as WDM, TDM, and OFDM
- Cascadability of several SOAs for intensity modulated signals, advanced modulation formats, polarization multiplexed signals, and radio over fiber signals.

In a converged access network, a 100 Gbit/s mm-wave wireless link may serve as a suitable fiber substitute, if fiber infrastructure is not available. The required bandwidth is found at high mm-wave frequencies between 200…300 GHz. The high-capacity wireless link supports intensity modulation and advanced modulation formats and can seamlessly be integrated into a fiber infrastructure.

2 Fundamentals

This *Chapter* gives fundamentals and basics, which are helpful for the understanding of this thesis. We briefly discuss inphase-quadrature (IQ) modulation (*Section 2.1*), typical transmitter (*Section 2.2*) and receiver (*Section 2.3*) implementations, semiconductor optical amplifier (SOA) basics (*Section 2.4*), and mm-wave electronics and photonics (*Section 2.5* and *2.6*).

2.1 Communication Signals and Systems

The purpose of a communication system is transmitting data over a physical medium, the communication channel. Fig. 2.1 depicts the basic elements of a communication system. At the transmitter (Tx), a modulator represents the interface to the communication channel. The aim of modulation is to upconvert the low-frequency baseband data from the data source to a bandpass signal at a (higher) carrier frequency which is can be transmitted over the physical medium. The modulated signal is sent from the transmitter over the communication channel to the receiver (Rx) where a demodulator recovers the baseband data from the received signal. In this *Section*, we discuss some general aspects of such a communication system which are relevant in the context of this thesis.

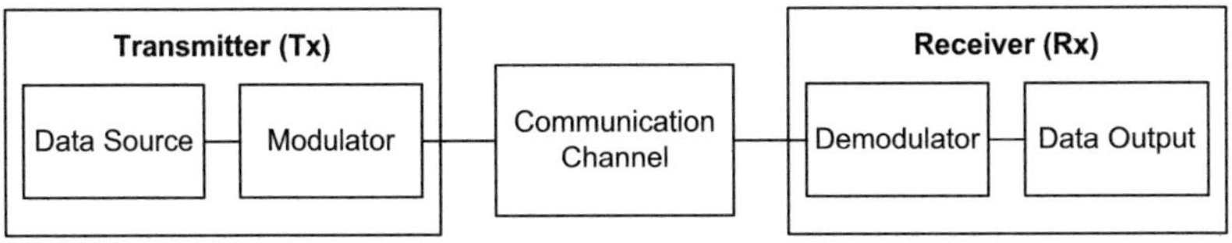

Fig. 2.1: Elements of a communication system: transmitter (Tx) with data source and modulation, communication channel, and receiver (Rx) with data demodulation and data output.

2.1.1 Communication Channel

The Shannon-Hartley theorem [75] describes the amount of data that can be transmitted over the communication channel. The theorem states that for an average transmitted signal power P the channel capacity C of a bandlimited channel of bandwidth B in the presence of additive white Gaussian noise (AWGN) with power spectral density ρ_{AWGN} is given by

$$C = B\log_2\left(1+\frac{P}{B\rho_{\mathrm{AWGN}}}\right)\ \text{(bits/s)}. \tag{2.1}$$

The ratio $P/(B\rho_{\mathrm{AWGN}})$ is called the signal-to-noise ratio (SNR). For a given channel bandwidth B, the channel capacity C may be increased without limit when $\mathrm{SNR}\rightarrow\infty$. However, for a given signal power P and no bandwidth limitation ($B\rightarrow\infty$), C does not become infinite, since also the noise power $B\rho_{\mathrm{AWGN}}$ increases. It can be shown, that for a communication system without bandwidth limitation, the maximum channel capacity is

$$C_\infty = \lim_{B\to\infty} C = \frac{P}{\rho_{\mathrm{AWGN}}}\log_2(e) = 1.44\frac{P}{\rho_{\mathrm{AWGN}}} \tag{2.2}$$

Besides noise, also other kind of signal degradations such as signal attenuation, nonlinear distortions, or multipath propagation may occur when transmitting data signal over a communication channel. Generally, the bandwidth of communication system is limited due to the physical limitations of the medium and the components which are used to build the transmitter and receiver. Typically, the amount of available bandwidth increases with the carrier frequency of the electromagnetic signal which carries the data.

Wireline Communication Channel

Traditional wireline channels are copper wire connections or coaxial cables. Copper wire lines in legacy telephone networks offer small bandwidths < 1 MHz. Twisted-pair cables used in Ethernet systems have usable bandwidths of several hundred MHz and allow for signal transmission over up to 100 m. Coaxial cables which are used for high-speed radiofrequency (RF) connections offer a bandwidth of several GHz. However, in view of the relatively large attenuation (typically 5 dB/m at 40 GHz) of such coaxial cables, the transmission distance is limited to only a few meters.

Wireless Communication Channel

In wireless communications channels, the signal is radiated from an antenna to the propagation medium, e.g. air. The employed carrier frequencies range from several kHz up to 300 GHz. The majority of the electromagnetic spectrum used for wireless communications is regulated and bandwidth is a very limited resource. While for mobile communications (carrier frequencies around 1 GHz and 2 GHz) the bandwidth per channel is limited to several MHz, GHz channel bandwidth is available in the unlicensed 60 GHz band. At some carrier frequencies, the wireless transmission distance is significantly influenced by the atmospheric attenuation and weather conditions (see

Section 1.4). Other prevalent effects in wireless communication are multi-path propagation and signal fading. At very high frequencies (mm-wave frequencies) mostly line of sight (LOS) connections are used, since highly directional antennas are needed to overcome the atmospheric losses and the free space path loss.

Fiber-optic Communication Channel

Fiber-optic communication systems offer channel bandwidths of many THz. The transmission medium is a fused-silica (glass) fiber. It has a very small attenuation of only 0.2 dB/km and 0.35 dB/km in the two low-loss transmission windows at the wavelengths 1550 nm (193 THz) and 1310 nm (229 THz), respectively. Laser light is guided in the optical fiber core by total internal reflection. The fiber core is surrounded by the fiber cladding which has a slightly lower refractive index than the core. A standard optical fiber is the single mode fiber (SMF) which has a core diameter of 8 µm. The SMF supports the propagation of only one mode in each of the two orthogonal polarizations. Multi-mode fibers (MMF) with a larger core diameter (> 50 µm) support the propagation of many modes. Depending on the operation frequency, data rate, and signal power, data signals may be affected by chromatic dispersion, scattering, and nonlinear effects when transmitted over an optical fiber.

2.1.2 Inphase Quadrature Modulation

Communication systems use electromagnetic signals to transmit data from the Tx to the Rx. A modulator at the Tx performs an upconversion of a baseband data signal $s(t)$ to a bandpass signal $E(t)$ at the carrier frequency f_0. For instance, f_0 may be in the order of some GHz for wireless communications or in the order of several THz for fiber optic communications. We assume a complex-valued baseband signal $s(t)$ with real part $s_\mathrm{I}(t)$ and imaginary part $s_\mathrm{Q}(t)$

$$s(t) = s_\mathrm{I}(t) + \mathrm{j}s_\mathrm{Q}(t) = A(t)\exp\left(\mathrm{j}\,\theta(t)\right) \tag{2.3}$$

where

$$\begin{aligned} A(t) &= \sqrt{s_\mathrm{I}^{\,2}(t) + s_\mathrm{Q}^{\,2}(t)}, \quad \theta(t) = \arctan\frac{s_\mathrm{Q}(t)}{s_\mathrm{I}(t)} \\ s_\mathrm{I}(t) &= A(t)\cos(\theta(t)), \quad s_\mathrm{Q}(t) = A(t)\sin(\theta(t)) \end{aligned} \quad . \tag{2.4}$$

The components $s_\mathrm{I}(t)$ and $s_\mathrm{Q}(t)$ are called the inphase and quadrature components. In a real system, one separate channel is needed for each of the components $s_\mathrm{I}(t)$ and $s_\mathrm{Q}(t)$ to represent the complex baseband signal $s(t)$. In general, I and Q data are independent.

By upconverting the baseband signal $s(t)$ to the carrier frequency f_0 we obtain the bandpass signal $E(t)$ which is real-valued (only one channel is needed to transmit the bandpass signal):

$$\begin{aligned} E(t) &= \Re\left\{s(t)\exp\left(\mathrm{j}2\pi f_0 t\right)\right\} = \Re\left\{\left(s_\mathrm{I}(t)+\mathrm{j}s_\mathrm{Q}(t)\right)\exp\left(\mathrm{j}2\pi f_0 t\right)\right\} \\ &= s_\mathrm{I}(t)\cos(2\pi f_0 t) - s_\mathrm{Q}(t)\sin(2\pi f_0 t) \\ &= A(t)\cos(\theta(t))\cos(2\pi f_0 t) - A(t)\sin(\theta(t))\sin(2\pi f_0 t) \\ &= A(t)\cos(2\pi f_0 t + \theta(t)) \end{aligned} \tag{2.5}$$

From the second line of Eq. (2.5) we see that the I component $s_\mathrm{I}(t)$ and the Q component $s_\mathrm{Q}(t)$ of the complex baseband signal modulate the amplitude of the two orthogonal carrier components $\cos(2\pi f_0 t)$ and $\sin(2\pi f_0 t)$. Using the relations of Eq. (2.4), we obtain a representation of the bandpass signal $E(t)$, which we may interpret as the electric field of an electromagnetic wave with amplitude $A(t)$, carrier frequency f_0, and phase $\theta(t)$. All three parameters can be used to encode data. However, in this thesis, we do not consider frequency modulation. The carrier frequency f_0 is kept constant.

Strictly speaking there is a fourth parameter which can be varied to encode data, namely the polarization. The polarization is not indicated in Eq. (2.5). For instance, in an optical communication system different states of polarization (SOP, e.g., *x*- and *y*- polarized light) can be used to directly encode data on the optical carrier (polarization shift keying, PolSK [76-78]). Two orthogonal SOPs may also be exploited in a so called polarization division multiplexing system (PDM) as two independent communication channels both employing IQ modulation [79, 80].

IQ Modulator

The basic implementation of an IQ modulator is shown in Fig. 2.2(a), corresponding to the second line of Eq. (2.5). The inphase baseband signal $s_\mathrm{I}(t)$ is multiplied (mixed) with a local oscillator (LO) $\cos(2\pi f_0 t)$. In the second arm of the LO path, a phase shifter creates a 90 degree phase-shifted version of the LO, which is mixed with the quadrature baseband signal $s_\mathrm{Q}(t)$. After subtracting the two mixer output signals, the composite real-valued bandpass signal $E(t)$ is transmitted.

An important aspect of IQ modulation is the fact that any arbitrary modulation of the amplitude $A(t)$ and the phase $\theta(t)$ of the bandpass signal $E(t)$ can be produced with the two baseband signals $s_\mathrm{I}(t)$ and $s_\mathrm{Q}(t)$. The implementation of the described IQ modulator is possible both in software and hardware. Hardware implementations of IQ modulators for the upconversion to optical carrier frequencies and to mm-wave frequencies are discussed in *Section 2.2* and *Appendix A.4.2*, respectively.

IQ Demodulator

Fig. 2.2(b) shows the basic implementation of an IQ demodulator at the Rx. The purpose of the IQ demodulator is to downconvert the received bandpass signal $E(t)$ to recover the complex-valued baseband signal $s(t)$.

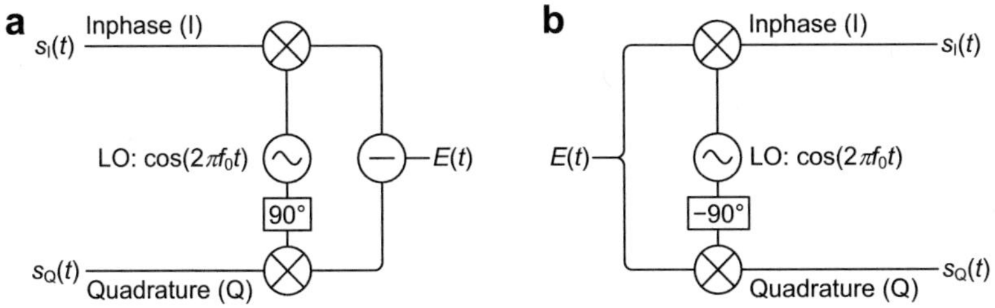

Fig. 2.2: Implementation of an IQ modulator (a) and demodulator (b). The IQ modulator up-converts the inphase and quadrature baseband signals $s_I(t)$ and $s_Q(t)$ to a composite bandpass signal $E(t)$ with carrier frequency f_0. The demodulator performs the inverse task. LO = local oscillator.

First, the received signal $E(t)$ is split in two arms. In the upper arm, the signal is multiplied with the LO $\cos(2\pi f_0 t)$ to recover the I component $s_I(t)$. In the lower arm, the signal is multiplied with the carrier $-\sin(2\pi f_0 t)$ to recover the Q component $s_Q(t)$. To correctly recover both the I and Q component, the LO at Rx must be phase-locked to the LO of the IQ modulator at the Tx. Therefore a carrier recovery (implemented either in hardware or in software) is needed at the Rx. Mathematically, the signal $s_{Rx}(t)$ after IQ demodulation can be written as

$$\begin{aligned} s_{Rx}(t) &= \Re\{s(t)\exp(\mathrm{j}2\pi f_0 t)\}\exp(-\mathrm{j}2\pi f_0 t) \\ &= \frac{1}{2}\left(s(t)\exp(\mathrm{j}2\pi f_0 t) + s^*(t)\exp(-\mathrm{j}2\pi f_0 t)\right)\exp(-\mathrm{j}2\pi f_0 t) \quad (2.6) \\ &= \frac{1}{2}s(t) + \frac{1}{2}s^*(t)\exp(-\mathrm{j}4\pi f_0 t) \end{aligned}$$

Besides the complex-valued baseband signal $s(t)$, the mixing process at the Rx also generates signal components at double the carrier frequency $(2f_0)$. In practice, these signal components are located outside the Rx bandwidth or filtered out using a low-pass filter. Hardware implementations of IQ de-modulators for the downconversion of optical carrier frequencies and mm-wave frequencies to the baseband are discussed in *Sections 2.3* and *A.4.2.*

2.1.3 Transmitter for Digitally Modulated Signals

In the previous *Subsection*, the concept of IQ modulation was introduced. In general, IQ modulation can be applied for both analog and digital modulation. Analog modulation techniques such as amplitude modulation (AM), frequency modulation (FM), or phase modulation (PM) are continuously variable. The signal is not encoded before transmission, i.e., the analog signal (e.g., a continuous voltage signal from a microphone) is simply modulated on a carrier that will propagate through the communication channel.

In communication systems, IQ modulation is especially advantageous when applied to digitally modulated signals. The terms "digital modulation" or "digitally modulated signals" describe the fact that digital data is mapped by a modulator to analog waveforms that match the characteristics of the communication channel. The modulation process is analog. However, the baseband data is in digital form. Compared to analog modulation, digitally modulated signals are spectrally more efficient, i.e., a larger amount of data can be transmitted in the same bandwidth. When using digitally modulated signals, one can also take full advantage of cost efficient digital signal processing (DSP) techniques to create and process the modulation.

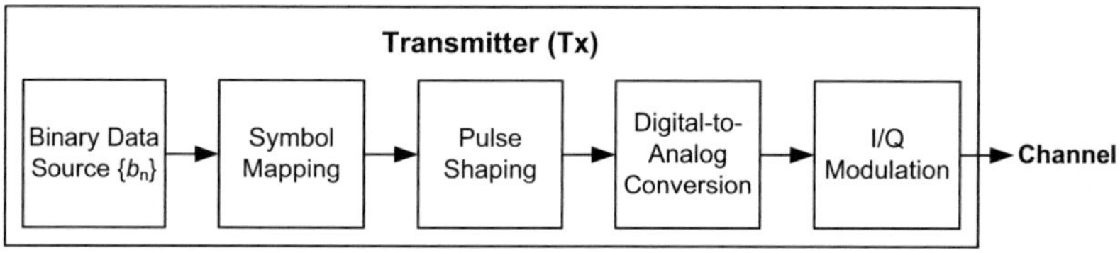

Fig. 2.3: Basic building blocks of a transmitter for generating digitally modulated signals: Binary data are mapped to (complex) symbols of a digital modulation format. Subsequent pulse shaping is used to improve the spectral efficiency of the signal. After digital-to-analog conversion (DAC), the inphase and quadrature components of the digital baseband signal are upconverted to the carrier frequency for transmission over the communication channel.

Fig. 2.3 shows a schematic of a Tx for generating digitally modulated signals. The data source provides a binary data sequence $\{b_n\}$ that needs to be mapped to analog waveforms before transmitted over the communication channel. In general, $\log_2 M$ bits from the data sequence $\{b_n\}$ are mapped at a time to one of M deterministic waveforms. The set of the M different waveforms is called modulation alphabet, and a waveform from this set is called symbol. The M symbols differ in amplitude, phase or a combination of both.

Constellation Diagram

A convenient representation of the different symbols is the constellation diagram which is basically an IQ diagram where the unmodulated carrier with carrier frequency f_0 is considered as the phase and frequency reference and the digitally modulated signal is interpreted relative to this reference. The inphase axis represents the 0 degree phase reference (I component), and the 90 degree out of phase quadrature axis represents the Q component.

Fig. 2.4(a) shows the quadratic constellation diagram with the ideal constellation points (positions of symbols) of a 16QAM signal (QAM = quadrature amplitude modulation). As can be seen, the 16QAM alphabet comprises $M = 16$ different symbols and each symbol represents a different sequence of $\log_2 16 = 4$ bits. For the shown 16QAM constellation, 3 amplitudes levels A_{mn} and 12 phase states θ_{mn} are combined.

The different symbol waveforms E_{mn} which are associated with the M different symbols of a quadratic QAM constellation may be expressed as

$$\begin{aligned} E_{mn} &= A_{mn} g(t)\cos(2\pi f_0 t + \theta_{mn}) \qquad , 0 \le t \le T_S \\ &= \Re\left\{\left(s_{\mathrm{I}m} + \mathrm{j} s_{\mathrm{Q}n}\right) g(t) \exp\left(\mathrm{j}2\pi f_0 t\right)\right\} \\ &= s_{\mathrm{I}m} g(t)\cos(2\pi f_0 t) - s_{\mathrm{Q}n} g(t)\sin(2\pi f_0 t) \end{aligned} \tag{2.7}$$

where

$$\begin{aligned} A_{mn} &= \sqrt{{s_{\mathrm{I}m}}^2 + {s_{\mathrm{Q}n}}^2} \\ \theta_{mn} &= \arctan \frac{s_{\mathrm{Q}n}}{s_{\mathrm{I}m}} \\ s_{\mathrm{I}m} &= (2m - 1 - \log_2 M), \qquad m = 1,2,\ldots,\log_2 M \\ s_{\mathrm{Q}n} &= (2n - 1 - \log_2 M), \qquad n = 1,2,\ldots,\log_2 M \end{aligned} \tag{2.8}$$

and T_S is the duration of one symbol period. The symbol rate $f_\mathrm{S} = 1/T_\mathrm{S}$. The amplitudes $s_{\mathrm{I}m}$ and $s_{\mathrm{Q}n}$ are the information-bearing inphase and quadrature components of the complex-valued baseband symbols. The real-valued waveform $g(t)$ in Eq. (2.7) is a pulse shaping function which is used to improve the spectral efficiency of the signal. As indicated in Fig. 2.3, pulse shaping is applied after the symbol mapping. The pulse shaping is discussed in *Section 2.1.5*.

Fig. 2.4(b) illustrates the mapping of a 20 bit-long binary sequence to the given 16QAM constellation of Fig. 2.4(a). The bit sequence $\{b_n\}$ is divided into blocks of $\log_2 16 = 4$ bits and mapped to the 5 symbols S1…S5. The amplitudes $s_{\mathrm{I}m}$ and $s_{\mathrm{Q}n}$ of the different complex symbols can directly be read from the I-axis and the Q-axis of the constellation diagram in Fig. 2.4(a). The first two diagrams in Fig. 2.4(b) show the amplitudes $s_{\mathrm{I}m}$ and $s_{\mathrm{Q}n}$ as a function of time. During one symbol duration T_{S}, the values of $s_{\mathrm{I}m}$ and $s_{\mathrm{Q}n}$ do not change.

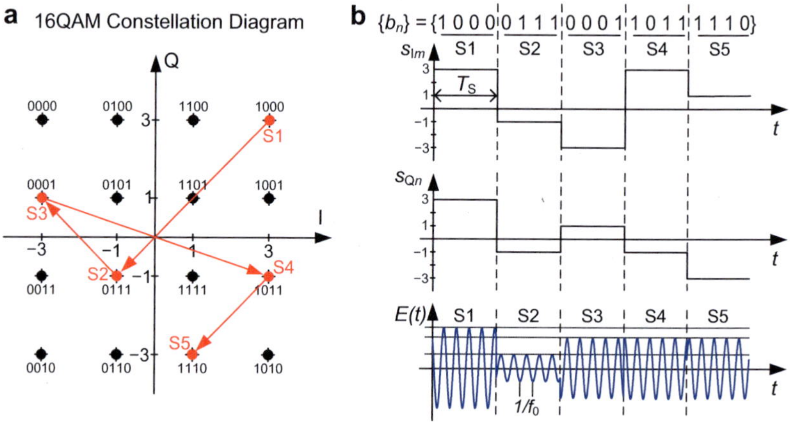

Fig. 2.4: (a) 16QAM constellation diagram (IQ plane) showing the positions and the associated binary data of each 16QAM symbol. The red symbols S1…S5 indicate the encoded symbols of the binary data sequence $\{b_n\}$ shown in (b). The arrows mark the transitions between two consecutive symbols. (b) Symbol mapping for the binary data $\{b_n\}$ and generation of the complex baseband symbols with inphase (s_{I}) and quadrature (s_{Q}) components according to the given 16QAM constellation in (a). The 16QAM signal after upconversion to the carrier with frequency f_0 is shown in the third diagram of (b).

A digital-to-analog converter (DAC), see Fig. 2.3, translates the digital baseband signals $s_{\mathrm{I}m}$ and $s_{\mathrm{Q}n}$ to analog waveforms (for both the I and the Q channel one DAC is needed). These waveforms are the input signals for the subsequent IQ modulator that upconverts the complex baseband signal to a real-valued bandpass signal $E(t)$ with carrier frequency f_0. The IQ modulation can either be done using a hardware IQ mixer, or with sufficient DSP resources, the IQ modulation may also be implemented in the digital domain (only a single DAC is needed).

The upconverted signal $E(t)$ of the previous 16QAM example is shown in the third diagram of Fig. 2.4(b). The three different amplitude levels of the upconverted 16QAM signal can clearly be distinguished. The symbols S3, S4, and S5 have the same amplitude level but different phase states. Note that for illustration purposes only a few carrier periods are drawn per symbol. In a real optical communication system, e.g., with $f_0 = 193$ THz and $f_S = 25$ GBd, there are about 7500 carrier periods per symbol (for smaller symbol rates f_S, the number of carrier periods per symbol even increases).

The red arrows in Fig. 2.4(a) indicate the transitions between the different symbols of the considered sequence given in Fig. 2.4(b). It can be seen, that the transition from symbol S1 to symbol S2 and the transition from symbol S3 to symbol S4 go through the origin of the constellation. This means that for a short period of time the carrier magnitude drops to 0. In optical communication systems, such changes may induce signal degradations when amplified with a semiconductor optical amplifier (SOA) [J5] see *Section 2.4.4*.

Besides 16QAM, many other digital modulation formats exist. At this point, we briefly introduce the modulation formats employed in the experiments of this thesis.

On-Off Keying

Fig. 2.5(a) shows the constellation of the on-off keying (OOK) modulation format which is the simplest digital modulation format as the binary data sequence is directly encoded by the presence ("on") or absence ("off") of the carrier. The OOK modulation format is widely used in optical communications where a binary data sequence signal is encoded in the intensity of the light. Neither IQ modulation nor IQ demodulation is needed for OOK. Since the OOK format is phase-insensitive it can easily be detected using square law detection (power detection $\propto |E(t)|^2$). Note that commonly the constellation diagram shows the electric field values (or electrical voltage values proportional to the electric field) and not power values.

Phase-Shift Keying and Quadrature Amplitude Modulation

In fiber-optic communications, modulation formats that go beyond the simple OOK modulation are widely denoted as advanced modulation formats [35]. Phase-shift keying (PSK) signals only modulate the carrier phase. Fig. 2.5(b) shows the constellation diagram of a binary (B)PSK signal. Like the OOK format, BPSK encodes 1 bit/symbol. In contrast to OOK, BPSK is a phase-sensitive format, where the data is encoded in the phase ($\pm\pi$) of a constant amplitude carrier (the BPSK format could also be interpreted as an amplitude-shift keying (ASK) format where the carrier changes its sign from

+1 to −1). No IQ modulator is required for the generation of BPSK signals, because the BPSK symbols are real-valued (all symbols are located on the inphase axis of the constellation diagram). Since the carrier magnitude of both BPSK symbols is equal, no simple square-law detection can be used at the Rx.

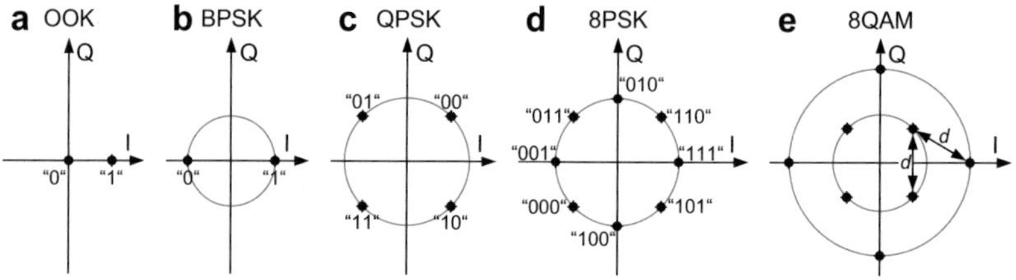

Fig. 2.5: Constellation diagrams of various digital modulation formats used in this thesis: (a) on-off keying (OOK), (b) binary phase shift keying (BPSK), (c) quadrature (Q)PSK, (d) 8PSK, and 8 quadrature amplitude modulation (QAM). The coding in (c,d) is Gray coding.

More common types of phase modulation are quadrature (Q)PSK, Fig. 2.5(c), and 8PSK, Fig. 2.5(d). A QPSK (8PSK) symbol consists of 2 bits (3 bits), which increases the spectral efficiency by a factor of 2 (3) compared to BPSK or OOK. It should be noted that although for PSK signals the carrier has the same constant amplitude for all symbols, crossings through the origin of the constellation diagram may occur during the transition between two consecutively modulated symbols. Then the carrier magnitude drops for a moment to 0. The zero crossings may be prevented by using variations of the presented digital modulation types such as Offset QPSK or constant-envelope modulation (Gaussian minimum shift keying) [81].

For encoding data with 3 bits/symbol, often circular 8QAM as depicted in Fig. 2.5(e) is preferred over 8PSK. The circular 8QAM constellation is known to be the best eight-point constellation, since it requires the least transmit power for a given minimum distance between the constellation points [82].

For most of the experiments conducted in this thesis, the different modulation formats were either generated with an arbitrary waveform generator (AWG) or with the software-defined multi-format transmitter from [83].

Gray Coding

There are many possibilities to map the binary data to the different symbols in a constellation diagram. The preferred mapping is the so called Gray encoding, where the binary representation of nearest adjacent symbols differs only by one bit. In the presence of AWGN, it is most likely that at the Rx a received symbol erroneously is mapped to one of its nearest neighbors in the constellation diagram. Since both symbols differ only by one bit, only a single bit error occurs in the $\log_2 M$ bits long sequence. The constellations shown in Fig. 2.4(a) and Fig. 2.5(c,d) use Gray coding.

2.1.4 Receiver for Digitally Modulated Signals

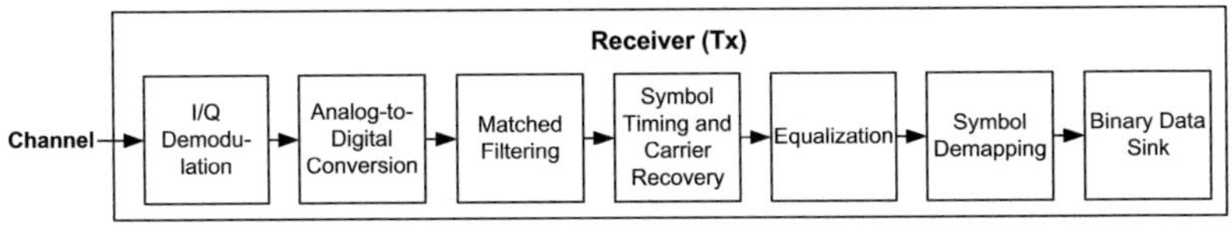

Fig. 2.6: Building blocks for receiving digitally modulated signals. After downconversion with an IQ demodulator, the inphase and quadrature components of the received signal are digitized using an analog-to-digital converter (ADC). In optical communication systems, chromatic dispersion compensation and polarization demultiplexing (both not shown in the Figure) are done after analog-to-digital conversion. Matched filtering is performed to minimize inter-symbol interference (ISI). Since the Rx local oscillator (LO) is not phase-locked to the Tx LO, symbol timing recovery and carrier recovery are needed at the Rx. A linear equalizer compensates for the frequency response of the communication system. Ideally, the original binary data from the Tx is obtained after symbol demapping.

Fig. 2.6 shows the schematic of a Rx for receiving digitally modulated signals. The Rx should fully recover the original binary data stream $\{b_n\}$ which was encoded at the Tx, see Fig. 2.3.

After transmission over the communication channel, the received bandpass signal is down-converted by an IQ demodulator to the baseband. In general, the Rx and Tx LO are not frequency locked (not to mention phase locked). This is the reason why both a carrier frequency recovery (carrier lock) and a symbol clock recovery (symbol lock) are necessary at the Rx to successfully recover the transmitted data. In modern communication systems these tasks are performed in the digital domain. Therefore the inphase and quadrature baseband data signals at the output of the IQ demodulator are digitized by an analog-to-digital converter (ADC). From this stage on, DSP is used for further processing and demodulation of the data. In an optical communication system, first the accumulated chromatic dispersion (CD) is compensated and polarization demultiplexing is performed for PDM systems (both not indicated in Fig. 2.6). In the next step, matched filtering (matched to the pulse shaping filter at the Tx) minimizes the inter-symbol interference (ISI). Signal pulse shaping and matched filtering are discussed in *Section 2.1.5*. After the matched filtering the carrier frequency, the carrier phase, and the symbol timing are estimated for carrier lock and symbol clock recovery [82]. A linear equalizer compensates the non-flat frequency response of the communication system. The baseband IQ waveforms are finally applied to a symbol detector that demaps the complex symbols and outputs a binary data stream, which in case of error-free transmission and demodulation is the original data stream $\{b_n\}$ encoded at the Tx.

The demodulated complex symbols may be compared with the ideal signal constellation diagram to determine error characteristics of the received signal such as the magnitude error, the phase error, and the error vector magnitude (EVM). The EVM metric (see *Appendix A.3.2*, [J6]) is a common performance measure for advanced modulation formats.

2.1.5 Nyquist Pulse Shaping

In the transmitter configuration of Fig. 2.3, a pulse shaper (filter) is applied to the complex baseband symbols. If the IQ modulator was driven by unfiltered square waves with fast transitions (as illustrated in the example of Fig. 2.4), then the bandpass signal would be spectrally very broad. In a regulated communication channel with many users, this might lead to problems like inter-channel interference. The pulse shaper in the Tx reduces the occupied signal bandwidth by slowing the fast transitions between the different symbols. However, the filtering causes overshoots and ringing effects that affect later symbols. In the worst case, ISI and a false interpretation of the received symbols at the Rx result.

Only pulses $h(t)$ that satisfy the Nyquist pulse shaping criterion lead to a transmission without ISI. The necessary and sufficient condition for a pulse $h(t)$ to satisfy the Nyquist pulse shaping criterion is that its Fourier transform $H(f)$ satisfies

$$\sum_{m=-\infty}^{\infty} H(f+\frac{m}{T_\mathrm{S}})=T_\mathrm{S} \tag{2.9}$$

A proof of the Nyquist pulse shaping criterion can be found in [81] or [82]. The condition stated in Eq. (2.9) is that the sum of $H(f)$ and its frequency shifted replicas is a constant. The pulse $h(t)=g_\mathrm{Tx}(t)*c(t)*g_\mathrm{Rx}(t)$ includes the pulse shaping at the Tx with $g_\mathrm{Tx}(t)$, the transmission over the communication channel having an equivalent lowpass impulse response $c(t)$, and the response $g_\mathrm{Rx}(t)$ of the Rx filter, i.e., $H(f)=G_\mathrm{Tx}(f)C(f)G_\mathrm{Rx}(f)$.

If we assume that the communication channel has a bandwidth of B, then $C(f)=0$ for $|f|>B$ and also $H(f)=0$ for $|f|>B$. When the symbol rate $1/T_\mathrm{S}>2B$, then the replicas of Eq. (2.9) do not overlap. The frequency spacing between two adjacent replicas is $1/T_S - 2B$. In this frequency spacing, Eq. (2.9) cannot be fulfilled, a system without ISI is not possible.

For the case $1/T_S = 2B$ (Nyquist rate), neighboring replicas of $H(f)$ touch each other and only $H(f) = T_S$ for $|f| < B$ satisfies Eq. (2.9). The corresponding pulse for this case is the sinc pulse

$$h(t) = T_S \frac{\sin(\pi t / T_S)}{\pi t} = \mathrm{sinc}\left(\frac{\pi t}{T_S}\right). \qquad (2.10)$$

This means that the largest symbol rate that can be transmitted without ISI is $1/T_S = 2B$. Sinc pulses mean perfect brick-wall filtering and an exact sampling is needed to avoid ISI.

When $1/T_S < 2B$, then the replicas of $H(f)$ are overlapping and many possible solutions for $H(f)$ exist. In practice, communication system use sinc-like pulses that satisfy the Nyquist pulse shaping criterion. These pulses have a raised cosine (RC) frequency characteristic $H_{RC}(f)$ with roll-off factors $0 \le \beta \le 1$

$$H_{RC}(f) = \begin{cases} T_S & , 0 \le |f| \le \frac{1-\beta}{2T_S} \\ \frac{T_S}{2}\left[1 + \cos\left(\frac{\pi T_S}{\beta}\left(|f| - \frac{1-\beta}{2T_S}\right)\right)\right] & , \frac{1-\beta}{2T_S} \le |f| \le \frac{1+\beta}{2T_S} \\ 0 & , |f| > \frac{1+\beta}{2T_S} \end{cases} \qquad (2.11)$$

The corresponding sinc-like pulse $h_{RC}(t)$ is expressed as

$$h_{RC}(t) = \mathrm{sinc}\left(\frac{\pi t}{T_S}\right) \frac{\cos(\pi \beta t / T_S)}{1 - (2\beta t / T_S)^2}. \qquad (2.12)$$

Fig. 2.7(a.b) shows examples of sinc-like pulses and their corresponding RC spectra for $\beta = 1$, $\beta = 0.35$, and $\beta = 0$. From Fig. 2.7(a) it can be seen that the pulses have a peak amplitude at $t = 0$ and are zero at all other multiples the symbol duration T_S. Therefore different symbols do not interfere.

The sharpness of the filter is described by the roll-off factor β, see Fig. 2.7(b). The roll-off factor β (also known as the excess bandwidth factor) is a measure of the occupied system bandwidth $B = (1+\beta)/T_S$. For $\beta = 0$ we have the brick-wall filter case, and for $\beta = 1$, the filter bandwidth is double the symbol rate ($2/T_S$). In (wireless) communication systems, typical roll-off factors are $\beta = 0.2 \ldots 0.5$.

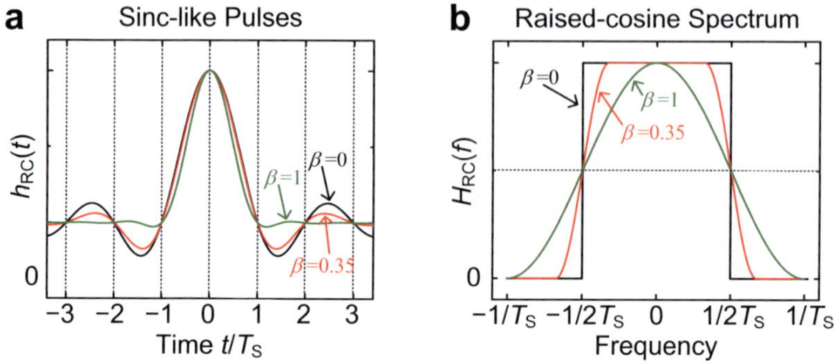

Fig. 2.7: Nyquist pulse shaping. Sinc-like pulses (a) and their raised-cosine spectra (b) for different roll-off factors β.

A single RC filter may be used at the Tx to achieve zero ISI. However, the filtering may also be shared between the Tx and the Rx. Then the Tx filter and the Rx filter must be closely matched to minimize ISI. If we assume an ideal communication channel with $C(f)=1$, we may evenly split the overall RC filtering by using a pair of matched root raised cosine (RRC) filters at the Tx and the Rx, i.e., $|G_{\text{Tx}}|=|G_{\text{Tx}}|=\sqrt{|H_{\text{RC}}|}$. Except for $\beta=0$, the impulse response of a RRC filter is not zero at multiples of T_{S}. It can be shown that matched RRC filtering maximizes the SNR at the Rx [81].

2.1.6 Orthogonal Frequency Division Multiplexing

Orthogonal frequency division multiplexing (OFDM) is a special kind of multicarrier transmission technique widely used in wireless communication systems [84-86]. Meanwhile, OFDM has found its way into optical communications [87-91],[J7], mainly enforced by the availability of low-cost high-speed DSP.

When data is transmitted simultaneously on different carriers over a communication channel, e.g., in a wireless frequency division multiplexing (FDM) system or in a fiber-optic wavelength division multiplexing (WDM) system, the different carriers must not overlap. In OFDM the subcarriers are orthogonal. Therefore the overlapping of the OFDM subcarriers does not impair the demodulation of signal at the Rx.

In the time domain, an OFDM signal is represented by a sequence of OFDM symbols. Each OFDM symbol is a superposition of N orthogonal sinusoidals with equidistant subcarrier frequencies f_k. The relation for the subcarrier frequency spacing $\Delta f = 1/T_{\text{sym}}$ follows from the symbol length

T_{sym} and the orthogonality condition of the subcarriers. A single complex-valued OFDM symbol $x(t)$ can be represented as

$$x(t)=\sum_{k=0}^{N}S_k\text{rect}\left(\frac{t}{T_{\text{sym}}}\right)\exp\left(\text{j}2\pi f_k t\right). \qquad (2.13)$$

The N OFDM subcarriers f_k carry different complex data S_k (e.g., QPSK or QAM symbols).

One major advantage of OFDM is the fact that modulation/multiplexing of the OFDM symbols at the Tx and demultiplexing/demodulation at the Rx can be implemented very efficiently in the digital domain by performing the inverse fast Fourier transform (IFFT) and the fast Fourier transform (FFT), respectively. In principle, Eq. (2.13) can directly be implemented in the digital domain employing the IFFT.

Since in OFDM data is transmitted in parallel on different subcarriers, a relatively long OFDM symbol length results compared to the symbol length in a serial data transmission system. Therefore, OFDM signals are much less affected by ISI. Usually, after each OFDM symbol an additional cyclic prefix (CP) is inserted, large enough to counteract the influence of the channel dispersion. For instance, a 25 % CP overhead means, that the first quarter of the OFDM symbol is copy-pasted to the end of the OFDM symbol. Introducing a CP reduces the net amount of transmitted data.

The frequency domain representation of an OFDM symbol as given by Eq. (2.13) is the sum of N sinc-functions centered at f_k. The disadvantage of this is that each OFDM subcarrier has sidelobes which extend over many other subcarriers. This makes OFDM very sensitive for frequency offsets and phase noise. Another disadvantage of OFDM is that it typically has a higher peak to average power ratio (PAPR) than single carrier modulation formats [92, 93] depending on the number of OFDM subcarriers and the modulation format used on each subcarrier. A high PAPR requires system components which have a wide dynamic range. For instance, the vertical resolution of a DAC may be fully consumed for generating an OFDM peak while the rest of the signal cannot be resolved sufficiently. In practice, the OFDM signal is clipped at the Tx to reduce the PAPR.

In this thesis, we investigate 5 GHz broadband, real-valued electrical OFDM signals that are transmitted over an optical fiber using intensity modulation. The OFDM signals are generated offline with the software MATLAB and then physically generated with a DAC.

2.2 Optical Transmitter

In the previous *Section*, the principle of signal modulation was introduced. In this *Section*, we briefly review typical implementations of optical transmitters also used in this thesis.

The most common and most simple technique of modulating an electric signal on an optical carrier (E/O conversion) is by *direct modulation*. Direct modulation means that the electric drive signal, e.g., the drive current of a laser diode, directly changes the laser light intensity. Directly modulated lasers (DML) are very compact, cost-efficient, and require only a few mA of drive current. Different types of DMLs exist, .e.g., Fabry-Perot diode lasers, distributed feedback lasers (DFB), or vertical cavity surface emitting lasers (VCSEL), each DML type having its particular advantages. However, despite the simplicity and low cost of DMLs, in general their main drawbacks are the limited bandwidth, the relative intensity noise, and the frequency chirp. With a DML, the optical intensity is modulated. Therefore only real-valued signals can be modulated. Complex digital baseband signals such as QPSK need prior upconversion to an intermediate frequency (IF).

In contrast to direct modulation, *external modulation* uses a separate device to modulate the intensity, the amplitude, or the phase of a continuous wave (CW) laser. The main advantages of external modulation over direct modulation are a larger modulation bandwidth, chirp-free modulation, and a precise control of the modulation process. One type of external modulator exploits the electro-absorption effect (Franz Keldysh effect or quantum confined Stark effect) in semiconductor materials to modulate the light. These so called electro-absorption modulators (EAM) offer very large modulation bandwidths. However, the operation of EAMs is wavelength dependent, and like DMLs, EAMs modulate the optical intensity.

Mach-Zehnder Modulator

The most common type of an external modulator is a lithium niobate (LiNbO3) modulator based on the linear electro-optic Pockels effect, where an external electric field (external drive voltage) changes the refractive index of the material. This leads to a phase change of the laser light. In a Mach-Zehnder modulator (MZM), see Fig. 2.8(a), the light $E_{\text{in}}(t)$ is split at the input in two arms. In each arm, the external voltages $v_1(t)$ and $v_2(t)$ induce a phase change $\vartheta_1(t)$ and $\vartheta_2(t)$, respectively. Depending on the phase relation of the light between both modulator arms, the output electric field $E_{\text{out}}(t)$ can vary between "off" (complete destructive interference) an "on" (constructive interference). In addition to the modulating voltages $v_{1,2}(t)$, a constant bias voltage v_{DC} is applied in one arm and induces a static phase shift $\vartheta_{\mathrm{DC}}(t)$ to set the desired operating point of the MZM. It should be noted that $LiNbO_3$ modulators are polarization sensitive devices and the correct SOP must be provided at the MZM input.

The field transfer characteristic T_{MZM} of the MZM is given by

$$T_{\mathrm{MZM}}=\frac{E_{\mathrm{out}}(t)}{E_{\mathrm{in}}(t)}=\exp\left(\mathrm{j}\frac{\vartheta_1(t)+\vartheta_2(t)}{2}+\mathrm{j}\frac{\vartheta_{\mathrm{DC}}}{2}\right)\cos\left(\frac{\vartheta_1(t)-\vartheta_2(t)}{2}+\frac{\vartheta_{\mathrm{DC}}}{2}\right). \quad (2.14)$$

Equation (2.14) indicates that there is both an amplitude modulation (cosine term) and a phase modulation (exponential function) of the light going through the MZM. In practice, a MZM is either used in push-push operation, i.e., $\vartheta_1(t)=\vartheta_2(t)$, which leads to a pure phase modulation, or in push-pull-operation, i.e., $\vartheta_1(t)=-\vartheta_2(t)$, which leads to a pure amplitude modulation.

In this thesis, the MZMs were used in push-pull operation. For push-pull operation, the field transfer characteristic of the MZM is given by

$$T_{\mathrm{MZM,push\text{-}pull}}=\frac{E_{\mathrm{out}}(t)}{E_{\mathrm{in}}(t)}=\exp\left(\mathrm{j}\frac{\vartheta_{\mathrm{DC}}}{2}\right)\cos\left(\vartheta_1(t)+\frac{\vartheta_{\mathrm{DC}}}{2}\right) \quad (2.15)$$

and squaring of Eq. (2.15) yields the intensity transfer function of the MZM

$$\left|T_{\mathrm{MZM,push\text{-}pull}}\right|^2=\cos^2\left(\vartheta_1(t)+\frac{\vartheta_{\mathrm{DC}}}{2}\right)=\frac{1}{2}\left[1+\cos(2\vartheta_1(t)+\vartheta_{\mathrm{DC}})\right]. \quad (2.16)$$

The static ($v_1(t)=0$ and thus $\vartheta_1(t)=\vartheta_2(t)$) field and the intensity transfer characteristic of the MZM for push-pull operation are shown in Fig. 2.8(b) as a function of the applied DC voltage. The voltage that is required for a π phase shift between both MZM arms is called the V_π voltage. As can be seen from Fig. 2.8(b), the operation point V_π (MZM null point) is needed for an amplitude modulation where the output optical field is proportional to the driving voltage $v_1(t)$. An intensity modulation (IM) that is quite proportional to the driving signal is achieved in a small region around the operating point $V_\pi/2$ (MZM quadrature point).

For generating OOK signals, the MZM is biased at its quadrature point and the two-level drive voltage varies between $+V_\pi/2$ and $-V_\pi/2$. If biased at the null point, BPSK can be generated by driving the MZM with a two-level signal varying between $+V_\pi$ and $-V_\pi$.

In general, a small V_π voltage is very desirable, because this enables small drive voltages and thus saves electrical driver amplifiers which would degrade the signal quality by adding electronic noise. Typical values of V_π voltages for $LiNbO_3$ modulators are in the order of 3…6 V. Recently, silicon organic hybrid modulators have gained interest [J4], because they may offer $V_\pi < 1$ V.

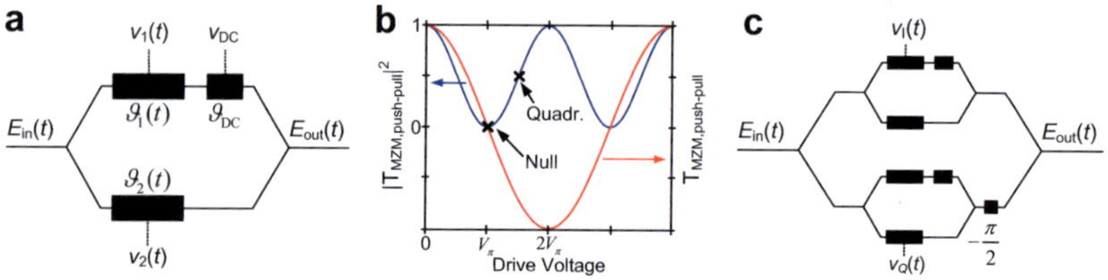

Fig. 2.8: (a) Mach-Zehnder modulator (MZM), (b) IQ modulator, (c) optical intensity and amplitude transfer function of a single MZM in push-pull operation. For amplitude modulation, the MZM is operated in the null point, for intensity modulation in the quadrature point (Quadr.).

Optical IQ Modulator

A single MZM as described above is not sufficient for generating advanced formats that use inphase and quadrature modulation. An optical IQ modulator that implements the IQ modulation functionality of Fig. 2.2(a) is shown in Fig. 2.8(c). The optical IQ modulator comprises two MZMs ("children") nested in a third MZM ("parent") structure. The child MZMs are operated at the null point (amplitude modulation) in push-pull mode. In the lower arm of the parent MZM, a −90° phase shift is induced with respect to the upper parent arm.

In this thesis, a software defined Tx (multi-format Tx) [83] comprising field-programmable gate arrays (FPGA) and DACs is used to generate electrical IQ signals which drive a commercial optical IQ modulator.

2.3 Optical Receiver

In optical communications, after transmission over the fiber, an opto-electronic (O/E) converter at the Rx converts the modulated light back into an electric current by means of photodetection. The different Rx configurations for detecting different kind of optical modulation formats are discussed in this *Section*.

2.3.1 Direct detection

For the detection of intensity modulated (IM) signals, square law detectors are used that produce a photocurrent $i(t)$ that is proportional to the mean squared optical electric field, i.e. a photocurrent that proportional to the incident optical power $p(t)$. This photodetection technique is called direct detection (DD) and commonly a single photodiode is employed. The generated photocurrent is given by

$$i(t) = Sp(t) \tag{2.17}$$

where S is the DC responsivity (in A/W) of the photodetector. The responsivity is a measure of the O/E conversion efficiency and is defined by

$$S = \eta \frac{e}{hf} = \eta \frac{e\lambda}{hc} \tag{2.18}$$

where η is the photodetector's quantum efficiency (limited to a theoretical maximum of 1), e is the electron charge, h is Planck's constant, f is the optical frequency, c is the speed of light in vacuum, and λ is the optical wavelength. An analysis of the signal-to-noise ratio (SNR) for DD with and without optical amplifier can be found in *Appendix A.2.1* and *A.2.3*.

The most commonly used photodetector for DD is the pin-photodiode (PD). A pin PD is a semiconductor junction diode with an intrinsic (i) layer between a p-doped and a n-doped layer. Typically, the structure is vertically illuminated. The light is absorbed in the intrinsic layer generating electron-hole pairs. By applying a reverse bias, the electron-hole pairs are swept by the electric field to the PD contacts where a photocurrent can be measured. There is a trade-off between the conversion efficiency and carrier transit time. The longer the absorption region of the PD the more light is absorbed and thus the quantum efficiency is increased. However, the time that the carriers need to reach the PD contacts also increases with the length of the absorption region which limits the bandwidth of the device.

By using edge coupling, the bandwidth of the PD can be increased. In an edge-coupled PD, the light directly enters the absorbing layer and propagates orthogonally to the applied electric field, i.e., orthogonally to the generated carriers [94]. This allows making the absorbing layer relatively long in the propagation direction of the light and at the same time designing a very thin absorption layer in the propagation direction of the carriers. Thus a large fraction of the light can be absorbed while the carrier transit time is small. However, one disadvantage of edge-coupled PD is the coupling efficiency between the PD and the coupling fiber.

Besides the carrier transit time, also the PD's RC time constant limits the bandwidth of the device. Especially for thin absorbing layers, the RC time constant becomes large. A comprehensive review of high frequency photodetectors is given in [95].

2.3.2 Coherent Reception

For coherent reception, the signal is mixed with a local oscillator (LO) at the photodetector. We write the electric field of the signal $E_S(t)$ as product of a slowly varying amplitude $A_S(t)$ and the optical carrier f_S (phase θ_S) [10]

$$E_S(t) = A_S(t)\exp(j2\pi f_S t + \theta_S) \tag{2.19}$$

For the optical electric field of the LO (frequency f_{LO} and phase θ_{LO} we write

$$E_{LO}(t) = A_{LO}(t)\exp(j2\pi f_{LO} t + \theta_{LO}) \tag{2.20}$$

The amplitudes $A_S(t)$ and $A_{LO}(t)$ are related to the respective signal power P_S and LO power P_{LO} by $P_S = |A_S|^2/2$ and $P_{LO} = |A_{LO}|^2/2$. In a coherent detection scheme, the LO is an inbuilt "amplification". Normally $P_{LO} >> P_S$ and thus high Rx sensitivities can be achieved even for small signal powers P_S.

Normally, balanced detection is used at a coherent Rx to suppress the DC components and maximize the signal current. For balanced detection, both the signal and the LO are combined using a 3 dB coupler which adds a 180° phase shift at one of the coupler's output ports. The signal from each output is detected with a different photodiode. Taking the difference of the photocurrents from both photodiodes eliminates the common direct current terms. As shown in *Appendix A.2.6*, the total output photocurrent of the balanced detector is given by

$$i(t) = 2S\sqrt{P_S P_{LO}}\cos(2\pi f_{IF} t + \theta_S(t) - \theta_{LO}(t)) \tag{2.21}$$

where $f_{IF} = f_S - f_{LO}$ is the so called intermediate frequency (IF).

Heterodyne Reception

The case where $|f_{IF}| > 2B$ (B is signal's baseband bandwidth) is called *heterodyne reception*. The signal received with heterodyne detection is a real-valued signal. However, a large bandwidth at the Rx is needed. Optical heterodyning on a sufficiently broadband photodetector, e.g. a uni-travelling carrier (UTC) PD as described in *Section 2.6.2*, may be used to convert (complex) data from an optical carrier to a wireless carrier at frequency $|f_{IF}| = |f_S - f_{LO}|$. An SNR analysis for optical heterodyne detection is given in *Appendix A.2.8*.

Homodyne Reception

The case $f_{IF} = 0$ is called *homodyne reception*. The photodiode current becomes

$$i(t) = 2S\sqrt{P_S P_{LO}}\cos\left(\theta_S(t) - \theta_{LO}(t)\right) \tag{2.22}$$

In the worst case, if $\theta_S(t) - \theta_{LO}(t) = \pi/2$, no signal is detected at all. This case can be avoided if the LO at the Rx is phase-locked to the signal carrier f_S. However, the implementation of an optical phase-locked loop (OPLL) is very complex. Furthermore, since only the signal's cosine (inphase) component is detected when employing coherent detection on a single photodetector, complex signals with inphase (I) *and* quadrature (Q) modulation such as QAM cannot be detected. For this purpose, an optical IQ demodulator which implements the IQ functionality of Fig. 2.2(b) is needed.

Optical IQ Demodulation with a 90° Optical Hybrid

In optical communications, the IQ demodulator functionality of Fig. 2.2(b) to detect IQ signals can be implemented using a 90° optical hybrid, see Fig. 2.9. A 90° optical hybrid is a six-port device. It comprises two 3 dB splitters for splitting both the signal (first input) and the LO (second input signal), a 90° phase shifter in one LO path (e.g., a quarter wave plate), and two 3 dB couplers. One 3 dB coupler combines the signal and the LO, the other coupler combines the signal and the 90°phase-shifted version of the LO. The two output signals of each 3 dB coupler are detected with the photodiodes of a balanced photodetector.

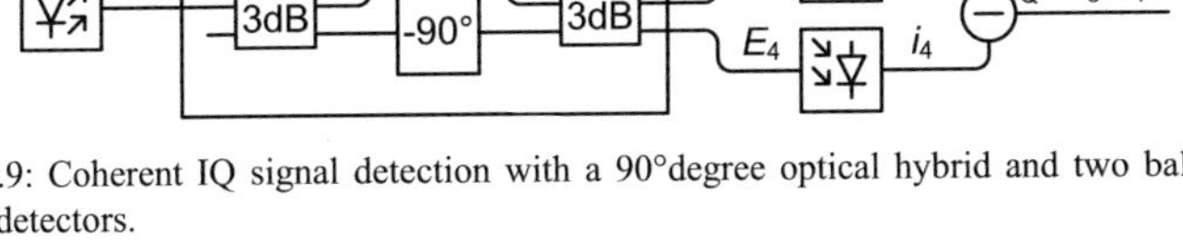

Fig. 2.9: Coherent IQ signal detection with a 90°degree optical hybrid and two balanced photodetectors.

When using phase-locked homodyne reception ($f_S = f_{LO}$, $\theta_{TX} = \theta_{LO}$), the output photocurrent $i_I(t)$ of one balanced detector directly represents the signal's I-component, and the photocurrent $i_Q(t)$ of the other balanced detector is the signal's Q-component. As shown in *Appendix A.2.7*, the complex baseband signal $i_C(t) = i_I(t) + \mathrm{j}i_Q(t)$ can then be restored.

In practice, frequency drifts between the carrier f_S (transmitter laser) and the free-running LO f_{LO} at the Rx cannot be avoided which results in a so called *intradyne reception* with $|f_{IF}| << B$. The frequency offset and phase estimation followed by carrier phase synchronization can be done with DSP [96, 97], which allows the use of standard DFB lasers in coherent optical communications [98]. For intradyne reception, a small amount of additional bandwidth compared to homodyne reception is needed.

For optical IQ demodulation with a 90° optical hybrid, the polarization of the incoming signal must be aligned to the polarization of the LO (indicated by the polarization controller in Fig. 2.9). However, in a real system polarization tracking and adjustment are not desirable and thus the SOP of the incoming signal is arbitrary. In the worst case, namely if the LO and the incoming signal have orthogonal SOPs, there is no output signal from the coherent receiver. In general optical coherent receivers are built in a polarization diversity configuration [10] where two separate 90° optical hybrids are used for the x and y polarization components. The incoming signal with arbitrary polarization and the linearly polarized LO are split into their x and y polarization components using a polarization beam splitter (PBS). The separated polarization components are then fed to the respective 90° optical hybrid. Not only does the polarization diversity scheme avoid polarization adjustment at the receiver, it also offers the possibility to receive polarization multiplexed signals, where each polarization carries independent data. The polarization demultiplexing can be done afterwards with DSP.

2.4 Semiconductor Optical Amplifier Basics

In this *Section*, we briefly introduce basic aspects and parameters of semiconductor optical amplifiers (SOA). The given information is useful for the understanding of the SOA related results presented in this thesis. For further information on SOA operating principles, different technological aspects, and detailed explanations on the underlying semiconductor physics, the interested reader is referred to [3, 99, 100].

2.4.1 Gain Characteristic of an SOA

An SOA is a semiconductor waveguide with a gain medium. In an SOA the gain is obtained by injecting carriers from a current source into the active region. These injected carriers occupy energy states in the conduction band of the active material leaving holes in the valence band. Electrons and holes recombine either non-radiatively or radiatively, in this case releasing the recombination energy in form of a photon. Three radiative processes are important in such structures, namely induced (= stimulated) absorption, spontaneous emission and induced (= stimulated) emission of photons [3]. For signal amplification, only the photons from induced emission are desired. However, the spontaneous emission of photons cannot be avoided in an SOA. The spontaneously emitted photons that are amplified in the active region of the SOA are the so-called amplified spontaneous emission noise (ASE).

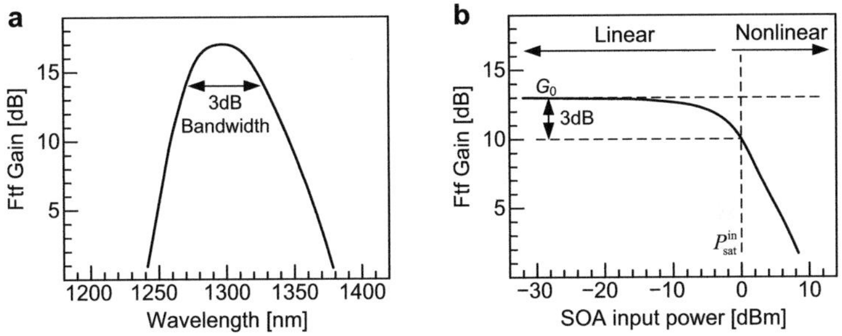

Fig. 2.10: Typical Ftf gain characteristic of an SOA. (a) The SOA bandwidth is around 60 nm with a peak gain of 17 dB at a wavelength of 1305 nm. (b) FtF gain as a function of the SOA input power. In this example, the unsaturated SOA gain G_0 is 13 dB. The input saturation power is 0 dBm and marks the transition from the linear to the nonlinear operating regime.

In a basic model, the optical power $P(z)$ at the propagation distance z in the active region of the SOA with the net modal gain g (unit 1/cm) can be described by

$$\frac{\mathrm{d}P(z)}{\mathrm{d}z} = gP(z) \tag{2.23}$$

The net modal gain g takes into account that the optical mode overlaps only partially with the active volume V. The net modal gain can be expressed by

$$g = \Gamma g_{\mathrm{m}} - \alpha \tag{2.24}$$

where Γ is the confinement factor (the fraction of the propagating power confined to the active region having an area A with width w and height d), g_{m} is the material gain coefficient, and α denotes the internal losses. The net modal gain g can be approximated by the differential gain a (unit cm^2), the carrier density N, and the carrier density at transparency N_{t}.

$$g \approx \frac{\mathrm{d}g}{\mathrm{d}N}(N - N_{\mathrm{t}}) = \Gamma a(N - N_{\mathrm{t}}), \qquad a = \frac{\mathrm{d}g}{\mathrm{d}N} \tag{2.25}$$

We assume that the net modal gain g is independent of the propagation distance z and that the waveguide has length L. Integration of Eq. (2.23) then gives the SOA chip output power $P(L)=P_{\text{out}}$ for a given SOA chip input power $P_{\text{in}}=P(0)$. We define the single-pass chip gain G of the SOA by

$$\begin{aligned} P(L) &= P(0)\exp\left(\int_0^L g\,\mathrm{d}z\right) = P(0)\exp(gL) \\ G &= \frac{P_{\text{out}}}{P_{\text{in}}} = \frac{P(L)}{P(0)} = \exp(gL) = \exp\left[a(N-N_{\text{t}})L\right] \end{aligned} \tag{2.26}$$

In practice, the fiber-to-fiber (FtF) gain G_{ff} is measured for packaged SOA chips. The FtF gain G_{ff} takes into account fiber-to-chip coupling losses at both chip facets and is smaller than the SOA chip gain G. If not stated differently, we refer to FtF gain values in this thesis and omit the index "ff". Fig. 2.10 shows schematically the typical FtF gain characteristic of an SOA as function of the wavelength (Fig. 2.10(a)) and as a function of the SOA input power Fig. 2.10(b). Commonly the SOA gain is given in dB values, $10\log_{10} G$. Today's SOAs have gain values in the order of 10…30 dB and 3 dB gain bandwidths of up to 100 nm.

Two SOA operating regions are distinguished, see Fig. 2.10(b). In the linear operating region (small input powers), the SOA gain is constant. Here the SOA output power is linearly proportional to the SOA input power (neglecting the SOA noise). In the linear region, signals are amplified with the unsaturated SOA gain G_0. The transition from the linear to the nonlinear SOA operating region is marked by the so-called saturation input power (Fig. 2.10(b)), which is the input power where the unsaturated SOA gain G_0 has decreased by 3 dB. (Sometimes also the input power for a 1 dB reduction of G_0 is taken). In the non-linear operating region, the output power increases sub-linearly and the SOA gain is decreasing. This gain suppression originates from the fact that the carrier injection rate which is fixed by the external current source sets a limit to the number of electron-hole pairs that can possibly recombine per time. Thus, the rate of generated photons is also limited [3].

To determine the SOA saturation input power, a simple carrier rate equation model can be used. The following derivation of the SOA saturation input power is taken from [3] (some intermediate steps were omitted). Equation (2.27) relates the temporal change of the carrier density N to the carrier injection rate I/e per active volume $V=AL=wdL$ and to the carrier recombination rates R_{c} and R'_{s}, which are due to spontaneous recombination (radiative and non-radiative), and to stimulated (radiative) transitions, respectively. The stimulated recombination rate due to the amplification of the signal is R_s, and the recombination rate due to ASE noise is R_{ASE}. The quantity τ_c is the so-called effective carrier lifetime, v_g the group velocity, and S' the total photon density. The total photon density S' consists of the signal photon density S and the ASE photon density S_{ASE}. Because of charge neutrality the carrier density N represents both, the electron and the hole concentration. Further, it needs to be mentioned that $v_g g = G'$ is the gain rate. The carrier rate equation reads

$$\frac{\mathrm{d}N}{\mathrm{d}t} = \frac{I}{eV} - R_c - R_s' = \frac{I}{eV} - \frac{N}{\tau_c} - g v_g S', \qquad R_s' = R_s + R_{ASE}, \quad S' = S + S_{ASE} \tag{2.27}$$

Using Eq. (2.25) and rewriting the total photon density S averaged over the active volume in terms of an average optical power $\Gamma P = S h f_s v_g A$, the relation (2.27) can be written as

$$\frac{\mathrm{d}N}{\mathrm{d}t} = \frac{I}{eV} - \frac{N}{\tau_c} - g v_g S = \frac{I}{eV} - \frac{N}{\tau_c} - \Gamma a (N - N_t) \frac{P}{A h f_s} \tag{2.28}$$

Solving Eq. (2.28) for a time-independent carrier concentration $\mathrm{d}N / \mathrm{d}t = 0$ leads to the following solution of Eq. (2.23)

$$\frac{P_{out}}{P_{in}} = G = G_0 \exp\left[-(G-1)\frac{P_{in}}{P_{sat}}\right], \qquad G_0 = \exp(g_0 L) \tag{2.29}$$

where the parameter P_{sat} is given by

$$P_{sat} = \frac{A h f_s}{a \Gamma \tau_c} \tag{2.30}$$

From Eq. (2.30), the input P_{sat}^{in} and output $P_{sat}^{out} = (G_0/2) P_{sat}^{in}$ chip saturation powers, for which the unsaturated single-pass chip gain is halved, can be calculated to be

$$P_{sat}^{in} = \frac{2\ln 2}{G_0 - 2} P_{sat} = \frac{2 h f_s \ln 2}{G_0 - 2} \frac{C}{\Gamma} \frac{1}{a} \frac{1}{\tau_c} \tag{2.31}$$

$$P_{sat}^{out} = \frac{G_0 \ln 2}{G_0 - 2} P_{sat} = \frac{h f_s \ln 2 G_0}{G_0 - 2} \frac{C}{\Gamma} \frac{1}{a} \frac{1}{\tau_c}\,. \tag{2.32}$$

2.4.2 Noise Figure of an SOA

The amount of ASE noise added by the SOA is described by its noise factor F, or the noise figure $\mathrm{NF} = 10\log_{10}(F)$. The noise factor of a single optical amplifier is derived in *Appendix A.2.4*. The noise factor is given by the inversion factor n_{sp} and the single-pass chip gain G

$$F \approx \frac{1}{G} + 2n_{sp}\frac{G-1}{G} \,. \tag{2.33}$$

The noise factor of an optical amplifier is minimized by using a population inversion factor n_{sp} approaching 1. The minimum achievable noise factor for an optical amplifier with a large $G >> 1$ is then $F = 2$ (or 3 dB).

The noise factor of a cascade of optical amplifiers is derived in detail in *Appendix A.2.5*. For the special case that the cascade consists of M identical SOAs and that the gain of an SOA compensates for the losses in the span after the SOA, the total noise factor of the cascade is given by $F_M = 1 + MF$.

2.4.3 Small-Signal Modulation Response of an SOA

In *Appendix A.1*, we use the rate equation given by Eq. (2.28) and a small perturbation calculation around an SOA input power operating point $P_{in,0}$ to derive the SOA small-signal frequency response $H(f)$. The small-signal SOA frequency response $H(f)$ is given by

$$H(f) = G\frac{\tau_2}{\tau_1}\frac{1 + j2\pi f \tau_1}{1 + j2\pi f \tau_2} \tag{2.34}$$

with the gain G from Eq. (2.29), and the two time constants τ_1 , and τ_2

$$\tau_1 = \tau_c\left(1 + \frac{P_{in,0}}{P_{sat}}\right)^{-1} \tag{2.35}$$

$$\tau_2 = \tau_c\left(1 + \frac{P_{in,0}}{P_{sat}}G\right)^{-1} \,. \tag{2.36}$$

The function $H(f)$ has a high-pass characteristic, since generally $\tau_1 > \tau_2$. With Eq. (2.31) we can rewrite Eq. (2.35) to

$$\tau_1 = \frac{\tau_c}{\left(1 + \dfrac{P_{in,0}2\ln 2}{P_{sat}^{in}\left(G_0 - 2\right)}\right)} \tag{2.37}$$

If $\left(G_0 - 2\right) >> 2\ln(2)P_{in,0}/P_{sat}^{in}$, then τ_1 corresponds to the effective carrier lifetime τ_c. In *Section* 3.2.2, we estimate the effective carrier lifetime τ_c from an SOA small-signal frequency response measurement. Note that the gain G in Eq. (2.34) is a function of the (optical carrier) frequency, see Fig. 2.10(a). However, for the small-signal response $H(f)$ we assume radio signals with a bandwidth much smaller (some GHz) than the optical carrier frequency (about 200 THz). Thus the gain G is considered to be frequency independent for the SOA small-signal response.

2.4.4 Signal Quality Limits when Amplifying *M*-ary QAM signals

The content of this Subsection is reprinted from the journal publication [J5]:

R. Bonk, G. Huber, T. Vallaitis, **S. Koenig**, R. Schmogrow, D. Hillerkuss, R. Brenot, F. Lelarge, G. H. Duan, S. Sygletos, C. Koos, W. Freude, and J. Leuthold, "Linear semiconductor optical amplifiers for amplification of advanced modulation formats," *Opt. Express*, vol. 20, no. 9, pp. 9657-9672, 2012, doi: 10.1364/OE.20.009657. © 2012 The Optical Society.

The quality of signals after an SOA is limited by SOA noise for low input powers [101], and by signal distortions due to gain saturation for large input powers [102]. Saturation of the gain not only induces amplitude errors but also phase errors due to the coupling of the alpha factor. Saturation also may induce inter-channel crosstalk if several wavelength division multiplexing (WDM) channels are simultaneously amplified by the same SOA. The ratio of the lower and upper power limits, inside which the reception is approximately error-free, is expressed by the input power dynamic range (IPDR), see *Subsection 2.4.5*.

Low Input Power Limit

For low input signal powers the limitations are due to ASE noise which in this case is virtually independent of the signal input power. Thus, if the input power decreases while the ASE power remains constant, the optical-signal-to-noise ratio (OSNR) will become poor. An example of such an OSNR limitation is presented in Fig. 2.11(a) for an optical QPSK (4QAM) signal. The constellation diagram shows a symmetrical broadening of the constellation points. The optimum situation where the input signal power is neither too low nor too high is shown in Fig. 2.11(b).

Large Input Power Limit

For large input powers the SOA gain is reduced due to gain saturation. Transitions between symbols are affected by the complex SOA response. Therefore, depending on the modulation format both the amplitude and phase fidelity of the amplification process are impaired to a different degree.

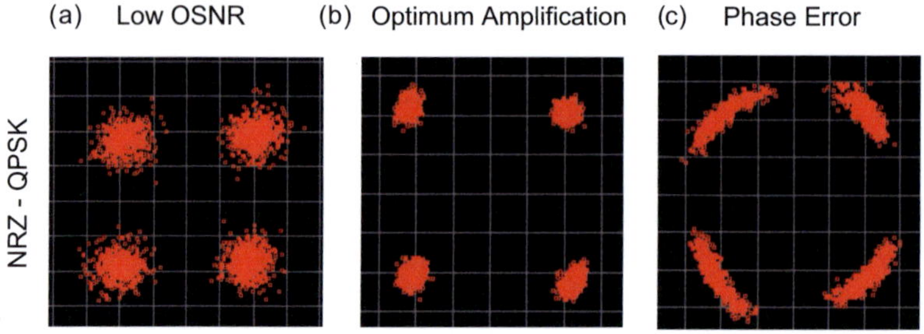

Fig. 2.11: Constellation diagrams showing the limitations of the signal quality for amplification of a 4QAM (QPSK) data signal. (a) The input power into the SOA is very low resulting in a low OSNR at the output of the amplifier. The constellation diagram of the 4QAM signal shows a symmetrical broadening of the constellation points. This broadening due to ASE noise causes a low signal quality. (b) Error-free amplification of the data signal is observed for a non-saturating input power. (c) For high input powers a nonlinear phase change induced by a refractive index change within the SOA causes a rotation of the constellation points. This rotation causes a reduction of the signal quality. Reprinted from [J5] © 2012 The Optical Society.

Among the many implementations of *M*-PSK and *M*-ary QAM formats the best performing transmitters often use zero-crossing field strength transitions [103], and therefore generate power transitions (solid line), see Fig. 2.12(a,b). These power transitions change the carrier concentration N and therefore the SOA fiber-to-fiber (FtF) gain $G_{\mathrm{ff}}=\exp\left(g_{\mathrm{ff}}L\right)$, where the FtF net modal gain g_{ff} is assumed to be independent of the SOA length L and comprises the SOA net modal chip gain g ($G=\exp\left(gL\right)$) as well as any coupling losses $\alpha_{\mathrm{Coupling}}$ to an external fiber, $G_{\mathrm{ff}}=\alpha_{\mathrm{Coupling}}G\alpha_{\mathrm{Coupling}}$, with $1>\alpha_{\mathrm{Coupling}}\geq 0$. A change $\Delta g_{\mathrm{ff}}=\Delta\left(\ln G_{\mathrm{ff}}\right)/L=\Delta g$ of the FtF net modal gain is identical to a change Δg of the net modal gain leading to a change Δn_{eff} of the effective refractive index n_{eff} by amplitude-phase coupling, which in turn is described by the so-called Henry's alpha factor α_{H}. With the vacuum wave number $k_0=2\pi/\lambda_{\mathrm{s}}$, signal wavelength λ_{s}, the complex output field is in proportion to $(G_{\mathrm{ff}})^{0.5}\exp\left(-\mathrm{j}k_0 n_{\mathrm{eff}}L\right)$ where the output phase (not regarding any input phase modulation) is defined by $\varphi=-k_0 n_{\mathrm{eff}}L$. Output phase change $\Delta\varphi$ and effective refractive index change are related by $\Delta\varphi=-k_0\Delta n_{\mathrm{eff}}L$. For the alpha factor we then find

$$\alpha_{\mathrm{H}}=\frac{\partial n_{\mathrm{eff}}/\partial N}{\frac{1}{-2k_0}\partial g_{\mathrm{ff}}/\partial N}\approx -2k_0\frac{\Delta n_{\mathrm{eff}}}{\Delta g_{\mathrm{ff}}}=\frac{2\Delta\varphi}{\Delta g_{\mathrm{ff}}L}=\frac{2\Delta\varphi}{\Delta\left(\ln G_{\mathrm{ff}}\right)}=\frac{2\Delta\varphi}{\Delta\left(\ln G\right)} \quad (2.38)$$

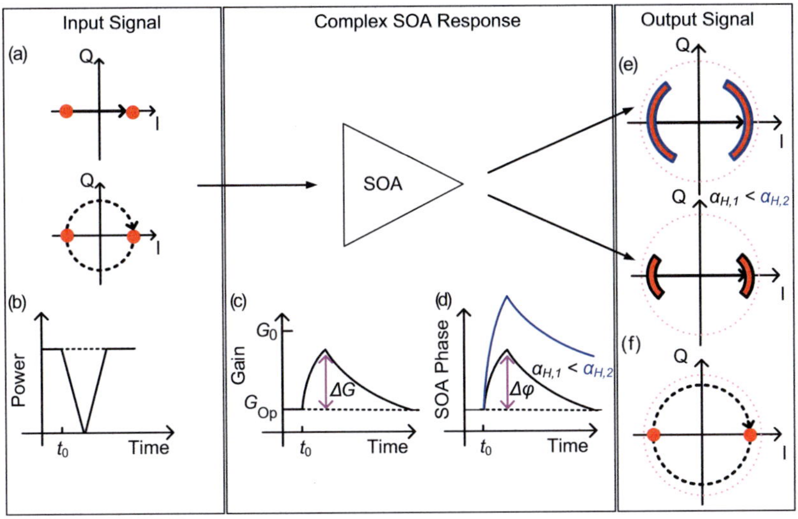

Fig. 2.12: Response of a saturated SOA in reaction to a BPSK (2QAM) signal with two possible transitions from symbol to symbol. Phase errors induced by power transitions from one BPSK constellation point to the other. (a) BPSK constellation diagram with in-phase (I) and quadrature component (Q) of the electric field. Solid line: zero-crossing transition; dashed line: constant-envelope transition. (b) Time dependencies for the two types of power transitions. SOA response that affects the (c) gain and (d) refractive index which leads to an SOA-induced phase deviation $\Delta\varphi$. SOAs with lower alpha factors induce less amplitude-to-phase conversion and therefore amplify the electric input field with a better phase fidelity. BPSK constellation diagram after amplification with a saturated SOA for (e) zero-crossing transition (for two alpha factors) and (f) constant-envelope transition. Reprinted from [J5] © 2012 The Optical Society.

Thus, by amplitude-phase coupling in the SOA, gain changes induce unwanted phase deviations. An illustration of this effect is schematically depicted in Fig. 2.12 assuming a BPSK format and a saturated SOA. If the signal power reduces at time t_0, Fig. 2.12(b), the gain starts recovering from its operating point described by a saturated chip gain G_{Op} (given by the average input power) towards the unsaturated small-signal chip gain G_0. After traversing the constellation zero the signal power increases and reduces the gain towards its saturated value G_{Op}, Fig. 2.12(c). Coupled by the alpha factor, a gain change induces a refractive index change and therefore an SOA-induced phase shift $\Delta\varphi$. SOAs with a lower alpha factor ($\alpha_{\mathrm{H,1}} < \alpha_{\mathrm{H,2}}$) have less amplitude-to-phase conversion and therefore give rise to less phase changes, see Fig. 2.12(d), and also Fig. 2.11(c). As a consequence, SOAs with lower alpha factors are expected to show better signal qualities for phase encoded data with a high probability of large power transitions. The constellation diagrams with induced phase errors are presented in Fig. 2.12(e) for the two alpha factors. If the transition between the two constellation points maintains a constant envelope (Fig. 2.12(a,b), dashed line) so that gain and phase changes do not happen (Fig. 2.12(c,d)), no phase errors occur, Fig. 2.12(f). If the device could be operated in the small-signal gain region even for high SOA input power levels, the constellation points would lie on the dotted circle. However, due to the described operation under gain saturation, the amplitudes at the SOA output are reduced, see Fig. 2.12(e,f).

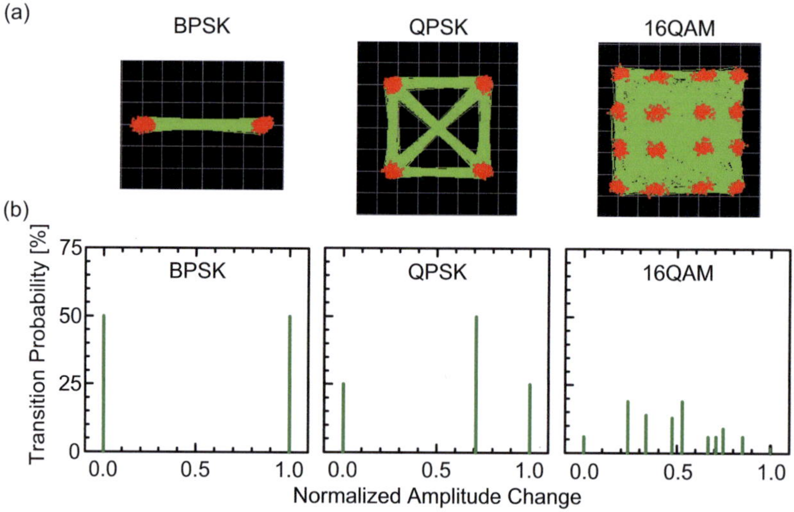

Fig. 2.13: Constellation diagrams and transition probabilities for different modulation formats. (a) Constellation diagrams with amplitude and phase transitions for BPSK, QPSK and 16QAM data signals. (b) Transition probability as a function of the amplitude change normalized to the largest possible amplitude transition. The probability of large transitions decreases for higher order formats. Reprinted from [J5] © 2012 The Optical Society.

Typically, the phase recovery in SOAs is slower than the gain recovery [104, 105], so a phase change induced at the power transition time has not necessarily died out at the time of signal decision (usually in the center of symbol time slot), so that the data phase is perturbed and errors occur. Additionally, the strength of the phase change induced by the SOA and detected at the decision point depends on the required time to change between constellation points, and thus on the symbol rate and the transmitter bandwidth. If the time to change between constellation points (approximately 10 ps…30 ps for

30 GBd…10 GBd) is not significantly faster than the SOA phase recovery time (about 100 ps), severe phase changes will degrade the signal quality. Thus, only at very high symbol rates, no significant phase change from the SOA is expected indicating the benefit of a limited SOA recovery time.

In addition to phase errors amplitude errors are expected to occur in *M-ary* QAM signals. Because *M*-ary QAM signals comprise multiple symbols with different amplitude levels extreme transitions from one corner to the other are less likely. Thus, phase errors due to amplitude-phase coupling are less likely as well. In average the amplitude distances between symbols reduce due to gain saturation.

Some examples of amplitude transitions in constellation diagrams for practical PSK and *M*-ary QAM implementations are shown in Fig. 2.13(a) for BPSK, QPSK and 16QAM. The transition probabilities for all occurring normalized amplitude changes are depicted in Fig. 2.13(b). The transition probability of the largest amplitude change reduces from 50% for BPSK to 25% at QPSK down to below 5% for 16QAM. Thus, the probability to observe a large amplitude change decreases the higher the order of the modulation format.

2.4.5 Dynamic Range of an SOA

The dynamic range of an SOA is an important parameter when deployed in a network. Here we introduce the (optical) input power dynamic range (IPDR) and the electrical power dynamic range (EPDR) of an SOA. Both quantities are used in this thesis to evaluate the SOA performance for different types of data signals and multiplexing techniques.

An SOA needs to amplify signals with both small and large optical input power levels. Thus an SOA with a large IPDR is required. The IPDR defines the power range in which error-free signal amplification can be achieved. The definition of error-free depends on the application and on the amount of signal overhead that is spent for error correction.

In the example shown in Fig. 2.14(a), the error vector magnitude (EVM) of a data signal was measured as a function of the SOA input power. The measured EVM can be related to the bit error ratio (BER) of the signal, see *Appendix* A.3.2 (however, note that in the nonlinear SOA operating region, the BER cannot be exactly calculated from the measured EVM as will be shown in *Section 3.1.3*). In the given example the IPDR is determined for a target BER of 10^{-9} which is indicated by the horizontal dashed line. The ratio of the input power limits $P_{\text{in},1}$ and $P_{\text{in},2}$ for error-free signal amplification (indicated by the vertical dashed lines) defines the IPDR measured in dB

$$\mathrm{IPDR} = 10\log\left(P_{\mathrm{in},2} / P_{\mathrm{in},1}\right) . \tag{2.39}$$

A large IPDR is obtained when the SOA has a large saturation input power $P_{\mathrm{sat}}^{\mathrm{in}}$ such that it can cope with largest amplitudes as well as when the SOA has a low noise figure such that it can deal with weak amplitudes [J5].

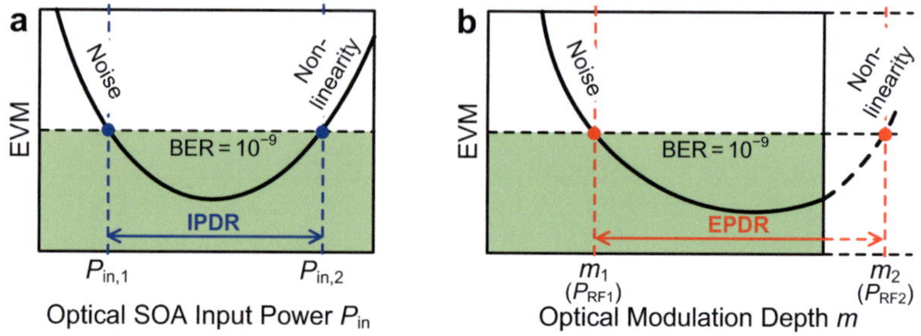

Fig. 2.14: (a) Input power dynamic range (IPDR) and (b) electrical power dynamic range (EPDR).

When amplifying radio-over-fiber (RoF) signals we use the term optical power dynamic range (OPDR) instead of IPDR. For the amplification of RoF signals, besides the OPDR also the EPDR of an SOA is important. The EPDR of an SOA is defined as the range of the optical modulation depth m in which error-free amplification can be achieved, see Fig. 2.14(b).

The optical modulation depth m is a measure of the RF signal power P_{RF} of the radio signal that is encoded on the optical carrier by intensity modulation. In the experiments of this thesis, a MZM is used to encode the radio signal on the optical carrier. If not stated differently in the text, we define the optical modulation depth as the ratio between the peak-to-peak signal voltage v_{pp} and the $V_{\pi,\mathrm{MZM}}$ voltage of the MZM

$$m = \frac{v_{\mathrm{pp}}}{V_{\pi,\mathrm{MZM}}} . \tag{2.40}$$

The ratio of the upper and lower limit of the optical modulation depth m_2 and m_1 for error-free signal amplification (m_2 and m_1 correspond to the RF power levels $P_{\mathrm{RF},2}$ and $P_{\mathrm{RF},1}$ of the encoded radio signal) defines the IPDR measured in dB

$$\mathrm{EPDR} = 20\log\left(m_2 / m_1\right) = 10\log\left(P_{\mathrm{RF},2} / P_{\mathrm{RF},1}\right) . \tag{2.41}$$

In practice, the upper limit of the EPDR (mainly determined by the nonlinearities of the electronic receiver) often lies at very high RF powers levels which are already beyond the measurement range. In such a case, the minimum guaranteed EPDR is given.

2.5 Mm-wave Photonics and Radio-over-Fiber

This *Section* gives a brief overview of microwave/mm-wave photonics and radio-over-fiber (RoF) technology. Comprehensive introductions to the field can be found in [106-108].

2.5.1 Microwave and mm-Wave Photonics

According to [106] microwave photonics comprises two aspects: "first, the study of opto-electronic devices and systems processing signals at microwave frequencies and, second, the use of opto-electronic devices and systems for signal handling in microwave systems."

Usually, the term microwave denotes the frequency range between some MHz and 30 GHz of the electromagnetic spectrum, and the term mm-wave is used when referring to the frequency range between 30 GHz and 300 GHz. The term photonics generally refers to frequencies in the infrared, visible, and ultraviolet part of the electromagnetic spectrum.

From the definition above, it becomes clear that microwave/mm-photonics combines the world of radiofrequency engineering and photonics. Initially, one major motivation of microwave photonics was the use of optical fiber as a transmission medium for microwave and mm-wave signals. Optical fiber is a very low-loss transmission medium over the entire microwave and mm-wave modulation frequency range and in addition, optical fiber offers a huge bandwidth. Further advantages compared to conventional electrical transmission systems with coax cables are reduced size, reduced weight, and immunity to electromagnetic interference. The weight and attenuation benefits of microwave/mm-wave photonic links over coaxial cables are compelling: 1.7 kg/km and 0.2 dB/km for fiber with 567 kg/km and 360 dB/km at 2 GHz for coaxial cable [107].

Fig. 2.15: Block diagram of a basic microwave/mm-wave photonics system.

The basic concept of a microwave/mm-wave photonics system is shown in Fig. 2.15. The RF signal is converted to the optical domain by electro-optic conversion (E/O). This is done either by direct or external modulation. Then, signal processing or signal transmission is performed in the optical domain. Finally, the signal is converted back to the electrical domain by opto-electronic conversion (O/E). In general, photodetectors with high responsivity, large bandwidth and high saturation current (i.e., large optical input power handling capability) are desirable. A specially designed device for mm-photonics is the so called uni-travelling carrier (UTC) PD that offers large electrical output powers and bandwidths even beyond 300 GHz. In this thesis, a UTC-PD is used to generate mm-wave signals around 240 GHz. More information on UTC-PDs can be found in *Appendix A.5*.

Typically, the E/O and O/E conversion introduce a loss in the order of 20…40 dB [109]. Therefore, transmitting RF signals over fiber makes only sense over distances where the overall loss is smaller than the loss in a coaxial cable. For mm-wave signals, this is already the case for transmission distances of a few meters. However, when using intensity modulation, fiber dispersion introduces a phase shift between the upper and lower sidebands, leading to a periodic signal power fading (fixed RF frequency) as a function of the transmission distance [110]. The larger the microwave/mm-wave frequency the shorter is the distance between two neighboring power drops. The influence of dispersion can be mitigated by dispersion-tolerant RoF transport schemes such as single sideband modulation [111-113].

In the following we introduce in more detail radio-over-fiber as a particular application of microwave/mm-wave photonics. Radio-over-fiber signals are of interest in this thesis, especially in *Section 3.2* and *Section 3.3* where RoF signals are amplified with an SOA. However, it should be pointed out that microwave/mm-wave photonics includes much more than RoF. In general, microwave/mm-wave photonics tries to take advantage of photonic technologies to implement functionalities that are difficult to implement directly in the RF domain. This includes, e.g., processing of microwave/mm-wave signals in the optical domain, filtering, analog-to-digital conversion, arbitrary waveform generation, or optical phased-array radars (see [107] for an overview).

2.5.2 Radio-over-Fiber

Radio-over-fiber (RoF) is one of the main applications of microwave/mm-wave photonics. In RoF systems, a fiber network/infrastructure is used for the transport and distribution of radio signals. Radio-over-fiber solutions are attractive for a variety of wireless systems, e.g., indoor distributed antenna systems (DAS) [114], cellular mobile networks, indoor wireless local area networks (WLAN), or fixed and mobile broadband networks [23].

As pointed out earlier, one major advantage of optical fiber as a transmission medium for radio signals is the small attenuation and the large bandwidth. This allows distributing radio signal over large distances to remote base stations (BS), where the radio signal is finally radiated from an antenna (sometimes the BS is denoted as remote antenna unit/site, RAU/RAS). In this kind of network architecture, the BSs are kept as simple as possible. Complex electronics and processing operations are located at the centralized central office (CO). This enables a cost-efficient implementation of a wide-area RoF network with many distributed, simple, and compact BSs. Ideally, the BSs are designed such that in case of a system upgrade only modifications are required at the CO. It should be noted that single-mode fiber (SMF) is the preferred transport medium for RoF signals in optical networks with distances of several km between the CO and the BS. For in-building networks, multi-mode fiber (MMF) is predominantly used worldwide, and RoF signals are transmitted over several hundred meters of MMF [115-118].

The increasing demand of bandwidth for wireless communications (see *Chapter 1*), leads to the use of higher carrier frequencies. An increase of the radio carrier frequency means, that the radio cell size reduces, thus a larger number of BSs is needed to cover the same area. Especially the license-free mm-wave frequencies in the 60 GHz band are of interest for high-capacity pico-cellular (indoor) wireless networks offering multi gigabit data rates per user. In a 60 GHz system, signal generation and distribution effectively can be done with mm-wave photonics and RoF technology [44, 119-124].

Different Signal Transport Schemes for RoF Systems

Fig. 2.16 shows different schemes for the RoF transport from the CO to the BS (downlink direction; for the uplink direction equivalent schemes can be applied). Usually intensity modulation and direct detection is used in all three schemes. The difference between the three schemes lies in the complexity of the BS and the CO, respectively.

The first scheme shown in Fig. 2.16(a) is called *RF-over-fiber* where the radio signal is transmitted over the fiber at the radio carrier frequency. At the CO, the baseband (BB) data signal is up-converted via an intermediate frequency to the radio carrier frequency and then translated to the optical domain by E/O conversion. The BS is very simple. Ideally, it only comprises an element for E/O conversion, if needed an RF amplifier, and an antenna. No further frequency up- or down-conversion of the signal is needed after photodetection. However, the RF-over-fiber scheme becomes challenging with high radio frequency carriers, since (costly) broadband optical modulation devices and broadband photodetectors are needed at the CO and BS, respectively. Furthermore, when using intensity modulation, chromatic fiber dispersion affects the RoF signal, as the RF power after photodetection is a function of the phase difference between the upper and the lower sideband of the RoF signal. The phase difference on the other hand depends on the fiber distance and fiber dispersion coefficient.

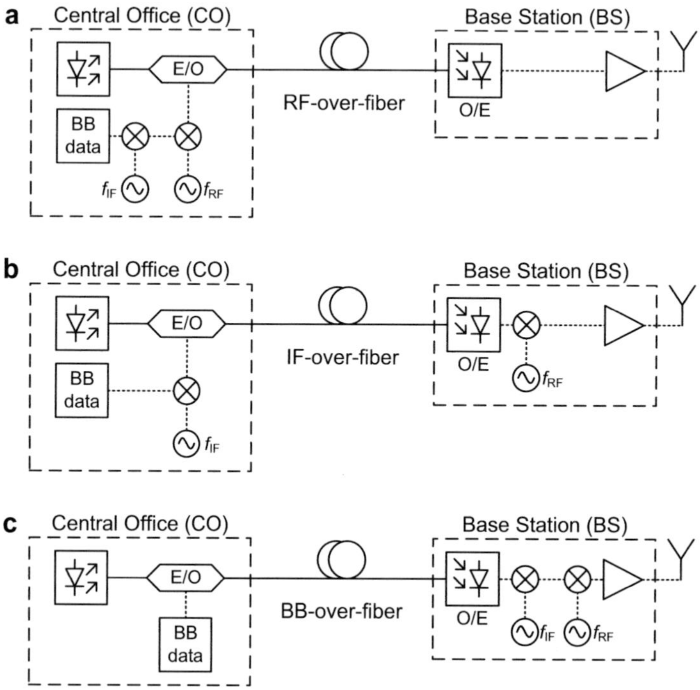

Fig. 2.16: Schematic diagram showing the hardware complexity for downlink transmission from the central office (CO) to the remote base station (BS) using the radiofrequency (RF)-over-fiber (a), intermediate frequency (IF)-over-fiber (b), or baseband (BB)-over-fiber (c) transport scheme.

The second and third schemes shown in Fig. 2.16(b,c) are called *IF-over-fiber* and *BB-over-fiber*, respectively. In these schemes, the complexity of the BS has increased compared to the RF-over-fiber scheme. Both schemes make use of additional, mature, low-cost electronics (LO, mixer) in the BS to upconvert the received IF signal (typical IF of 1 GHz) or BB signal after photodetection to the desired radio frequency. This relaxes the bandwidth requirements for the optical components in both the CS and the BSs. However, the additional and particular electronic hardware (mixers for a certain frequency band and a LO at a fixed frequency) that is needed for IF-over-fiber and BB-over-fiber make these schemes relatively inflexible for upgrades to new radio frequencies. Chromatic dispersion does not play a significant role in these schemes.

In this thesis, RF-over-fiber signals are investigated when amplified with an SOA in an access network scenario (*Section 3.2* and *Section 3.3*).

2.6 Combining mm-Wave Electronics and Photonics

In this thesis, a high-capacity mm-wave link is embedded in a fiber-optic environment. Two system concepts are investigated for the generation of mm-wave signals: electronic up-conversion and photonic up-conversion. For the frequency down-conversion after wireless transmission, electronic mixers are used. *Subsection 2.6.1* is dedicated to the electronic up- and down-conversion technique with electronic mixers. In *Subsection 2.6.2* the concept of photonic mm-wave generation by optical heterodyning is introduced.

2.6.1 Electronic Up- and Down-Conversion

Electronic Up- and Down-Conversion of Real-Valued Signals

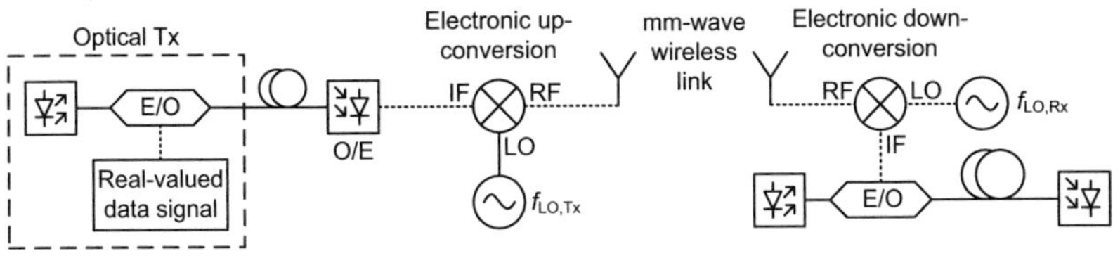

Fig. 2.17: Integration of a mm-wave wireless link in an fiber optic environment. The mm-wave signal is generated after photodetection by electronic up-conversion using active mm-wave monolithic integrated circuit (MMIC) mixers. After wireless transmission and electronic down-conversion to baseband, the signal can be further transmitted over optical fiber. In the shown configuration, the electronic mixers only have a single inphase IF signal port, thus only real-valued data signal can be transmitted.

The system concept based on electronic up-conversion for transmitting real-valued signals is shown in Fig. 2.17. At the optical transmitter, a real-valued data signal, e.g. an OOK signal, is encoded on an optical carrier and transmitted over an optical fiber. After direct detection with a photodetector the electrical baseband signal is up-converted to the mm-wave frequency carrier using an electronic mixer. The up-converted mm-wave signal is radiated from an antenna, transmitted wirelessly, and received with another antenna. After the receiving antenna, an electronic mixer down-converts the mm-wave signal to the baseband. The baseband signal may then again be encoded on an optical carrier for further transmission over optical fiber.

The integration of a mm-wave wireless link operating at a carrier frequency of 220 GHz into a fiber-optic infrastructure is presented in *Section 4.1* of this thesis. Additional information on the electronic mixer technology is given in *Appendix A.4.1*.

The Tx and Rx mixers only feature one IF port (inphase). This restricts the type of signals that can be transmitted and received with such a system configuration.

First, only real-valued signals can be transmitted, i.e., either a baseband signal that only has an inphase component or a complex signal that has been up-converted to an IF.

Second, the type of modulation that can be used depends on the Rx architecture (in this thesis, only active electronic mixers were used for down-conversion). In the following discussion we assume that we want to recover the baseband signal after wireless transmission:

If we modulate an OOK data signal with the two amplitude levels 0 and 1 at the Tx mixer (this requires DC coupling), the mm-wave carrier is absent for a sent 0, and the mm-wave carrier is present for the duration of the sent 1. We can use the Rx MMIC from Fig. A.5(b) for receiving the OOK signal. However, this requires a clock recovery device to lock the LO at the Rx to the mm-wave carrier frequency. Without clock recovery, the received signal would be zero if the Tx LO and the Rx LO are 180 degrees out of phase.

Clock recovery can be avoided by using a simple envelope detection scheme for example with a Schottky barrier diode (SBD), which rectifies the received mm-wave carrier.

Another technique to avoid the clock recovery for OOK signals is the use of an IQ-mixer at the Rx followed by the function $\sqrt{I^2+Q^2}$ which may be implemented in hardware or with DSP. This scheme requires DC coupling of the I and Q output signals at the Rx, since otherwise the information is lost when squaring the signals. However, in practice AC coupling is preferred, because thereby unwanted DC offsets that commonly occur in a direct down-conversion receiver [125] can be eliminated and also AC coupled RF amplifiers may be applied to the I and Q signal. It can be shown mathematically that an intradyne reception followed by the operation $\sqrt{I^2+Q^2}$ also recovers the OOK signal. We experimentally demonstrated this kind of concept with multi-gigabit data signals at 240 GHz, [C2]. In general, the LO leakage at the Tx and the IQ imbalance at the Rx should be as small as possible when using the $\sqrt{I^2 + Q^2}$ approach, because the LO leakage leads to an unwanted offset of the received 0 level and an IQ imbalance introduces oscillations on the 1 level. The IQ imbalance can be compensated by DSP [126, 127].

Clock recovery at the Rx MMIC is also mandatory if a BPSK signal with the amplitude levels −1 and +1 is applied to the Tx mixer. For the BPSK signal no envelope detection can be used at the Rx, since both amplitude levels would become indistinguishable. A BPSK signal is for example present at the input of the Tx mixer, if an OOK signal after photodetection is amplified with an AC-coupled RF amplifier.

Electronic Up- and Down- Conversion of IQ-Signals

For electronic up-/down-conversion of complex valued IQ signals to/from mm-wave frequencies electronic Tx and Rx MMIC IQ mixers are needed. In such a scenario, see Fig. 2.18, any kind of complex modulation format is applied to an optical IQ modulator and transmitted over optical fiber. The optical signal is received by an optical 90° hybrid (see *Section 2.3.2*) and the I and Q baseband signals drive the electronic Tx MMIC IQ mixer. After wireless transmission the signal is down-converted by an Rx MMIC IQ mixer. The received I and Q components drive an optical IQ modulator for further data transmission over an optical fiber.

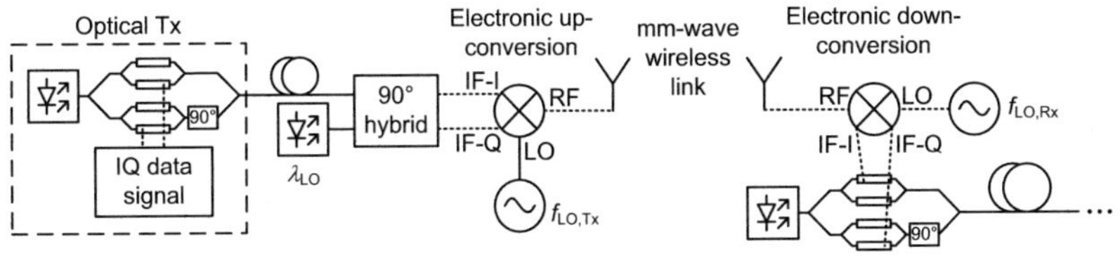

Fig. 2.18: Integration of a mm-wave wireless link in an fiber-optic environment with electronic up- and down-conversion of IQ data signals using a Tx MMIC IQ and a Rx MMIC IQ mixer, respectively.

Note that the I and Q signals at the output ports of the 90° optical hybrid and the Rx MMIC IQ mixer are not necessarily the original I and Q signal components that were encoded at the optical IQ Tx on the optical carrier. If the optical LO at the optical 90° hybrid is not locked to the optical carrier and the LO at the Rx MMIC mixer is not locked to the LO at the Tx MMIC, the constellation diagram is rotated. Thus the I and Q signals at the output ports of the optical 90° hybrid and the Rx MMIC IQ mixer, respectively, are different to the original I and Q signals. However, as mentioned in *Section 2.1.4*, carrier recovery and phase locking can finally be done with DSP.

Detailed information on the MMIC IQ mixers that were used for the experiments in this thesis is given in *Appendix A.4.2*. The Tx and Rx MMIC IQ mixers transmit and receive mm-wave signals at a carrier frequency around 240 GHz. In *Appendix A.4.2* it is shown that the Tx IQ mixer has a small linear dynamic range, which limits the operation to PSK signals with only one amplitude level. For achieving data rates of 100 Gbit/s and more, 16QAM modulation and linear system components are essential. In our experiments with IQ modulation, we replace the electronic Tx MMIC IQ mixer by a linear photomixer module, which enables the generation of a 100 Gbit/s 16QAM signal at a carrier frequency of 237.5 GHz (*Section 4.2*). For down-conversion we always rely on the electronic Rx MMIC IQ mixer module, which offers sufficient linearity.

2.6.2 Photonic mm-Wave Signal Generation

An elegant photonic technique for mm-wave signal generation is optical heterodyning (or photomixing). The principle of optical heterodyning was introduced in *Section 2.3.2*. The system concept when using photonic up-conversion is illustrated in Fig. 2.19. At the optical transmitter, a real-valued or complex-valued IQ data signal is encoded on the optical carrier f_1. The optical data signal is transmitted over an optical fiber and then superimposed with a LO (f_{LO}) on a photodetector. By square law detection, the mm-wave frequency $f_{\mathrm{RF}} = |f_1 - f_{\mathrm{LO}}|$ is generated (the sum frequency $f_1 + f_{\mathrm{LO}}$ is not generated in a photodetector). After wireless transmission, the signal is down-converted by a Rx MMIC IQ mixer (*Appendix A.4.2*). From this point, the scheme is identical to the one explained in Fig. 2.18.

There are several benefits of the photomixing technique: One is that the radio signal is generated remotely at the BS and that the signal can be transmitted and distributed over large distances over low-loss optical fiber. The required LO for photomixing may either be delivered together with the signal over the fiber, or provided locally as drawn in Fig. 2.19.

Another advantage is that the radio signal is directly generated from the optical signal without a detour to the baseband, i.e., neither an optical 90° hybrid nor an electronic IQ mixer is needed at the BS. If besides the optical data signal also the LO is delivered over fiber to the BS, the BS can be simplified to a photodetector and a radiating antenna.

Last but not least, the mm-wave carrier frequency f_{RF} can easily be tuned by tuning the frequency spacing between the two optical frequencies f_1 and

f_{LO}. A system for radiofrequency signal generation with over seven octaves of continuous tuning was achieved in [128] with a sideband-injection locking scheme. The maximum tuning range is mainly determined by the bandwidth of the photodetector. In this thesis, optical heterodyning in an ultra-broadband UTC-PD is used to generate mm-wave signals around 240 GHz. The operation principle of an UTC-PD is described *Appendix A.5*.

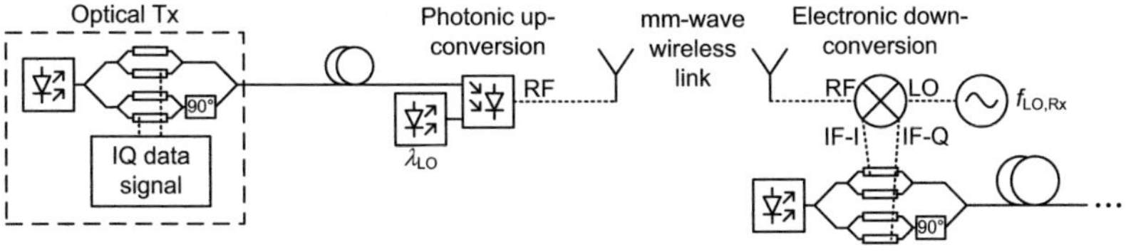

Fig. 2.19: Photonic mm-wave signal generation by optical heterodyning (photomixing). One advantage of this technique is that no detour to the baseband is necessary, i.e., neither an optical 90° hybrid nor an electronic IQ mixer is needed.

3 SOAs in Converged Access Networks

Semiconductor optical amplifiers (SOA) are considered to be key devices for compensating losses in future converged access networks because SOAs are inexpensive and offer a broad and variable gain spectrum. Various application scenarios with SOAs in the context of metro-access network convergence and optical-wireless network convergence have been introduced in *Section 1.3*. In this *Chapter*, we experimentally investigate specific aspects of these application scenarios. We identify limitations in dependence of the SOA operation conditions. The main aspects which are studied with respect to the amplification with SOAs are:

- *Different kind of data signals*: We investigate advanced optical modulation formats (*Section 3.1*) such as BPSK, QPSK, 8PSK or 16QAM which enable ultimate capacities > 100 Gbit/s in future access networks, e.g. in high capacity business paths between enterprises. Advanced modulation formats use the optical field and phase to encode the data and require coherent reception. We also study radio-over-fiber (RoF) signals (*Sections 3.2, Section 3.3*, and *Section 3.4*) which are of special interest in converged optical-wireless networks. For RoF signals, a small signal optical intensity modulation (IM) is employed at the transmitter and simple direct detection (DD) is performed at the receiver.
- *Multi-wavelength operation and different multiplexing techniques*: For advanced modulation formats we study the influence of multi-wavelength operation and polarization multiplexing (*Section 3.1*) enabling the transmission of aggregate data capacities in the order of Tbit/s. The amplification of multiple wavelengths is an important aspect in wavelength division multiplexing (WDM) scenarios, and polarization multiplexed 25 GBd QPSK signals constitute a kind of standard implementation of a 100 Gbit/s channel. For RoF applications, we investigate the amplification of broadband electrical orthogonal frequency division multiplexed (OFDM) signals (*Section 3.4*).
- *Simultaneous amplification of different kind of data and multiplexed signals*: Semiconductor optical amplifiers which act as a reach extender in future access networks need to cope with different kind of data and multiplexed signals at the same time. In *Sec-*

tion 3.3, we study the impact of a rival WDM signal, a rival bursty time division multiplexing (TDM) passive optical network (PON), and a rival RoF signal on a RoF subscriber in a converged RoF and WDM–TDM–GPON upstream scenario.

- *SOA cascades:* Converged metro-access ring networks rely on SOAs as loss compensating elements in reconfigurable optical add-drop multiplexers (ROADM). Data signals are transmitted over several ROADMs, thus a cascade of SOA results. We investigate both the amplification of advanced modulation formats (*Section 3.1*) and the amplification of broadband RoF OFDM signals (*Section 3.4*) by a cascade of several SOAs.

In this thesis we are especially interested in the power dynamic range that is offered by the SOA in the respective application scenarios and for the different kind of data signals. For optical signals we define the input power dynamic range (IPDR) as the power range in which error-free amplification is guaranteed. Error-free amplification is related to a certain bit error ratio (BER) limit, e.g. 10^{-3} or 10^{-9}. For RoF signals, we call the IPDR optical power dynamic range (OPDR) to distinguish it from the electrical power dynamic range (EPDR). The EPDR is defined in analogy to the OPDR as the range of the optical modulation depth of the RoF signal where error-free amplification is obtained (for a given optical SOA input power). Different modulation depths correspond to different power levels of the radio signal which is encoded on the optical carrier.

Of course, SOAs with large power dynamic ranges are desirable. Therefore a large upper and a small lower SOA input power limit are required. Linear SOAs with a constant gain over a large range of input powers (large saturation input power) and a small noise figure fulfill these needs. The SOAs investigated in this thesis are (linear) commercial QW SOAs and a (linear) QD SOA. Quantum dot SOAs are strongly polarization dependent (10 dB). However, they are expected to become polarization insensitive in the future [129]. We show that both types of SOA offer large power dynamic ranges and are well suited for the aforementioned applications scenarios. However, it should be noted that the explicit comparison of QW SOAs and QD SOAs is not the focus of this thesis. For such a comparison, the reader is referred to [3].

3.1 SOA Cascade with Advanced Modulation Formats

The content of this Section has been published in [J1]:

S. Koenig, R. Bonk, H. Schmuck, W. Poehlmann, Th. Pfeiffer, C, Koos, W. Freude, and J. Leuthold, "Amplification of advanced modulation formats with a semiconductor optical amplifier cascade," *Opt. Express*, vol. 22, no. 15, pp. 17854-17871, 2014, doi: 10.1364/OE.22.017854. © 2014 The Optical Society.

Parts of this Section have also been published in [C11]:

S. Koenig, R. Bonk, R. M. Schmogrow, A. Josten, D. Karnick, H. Schmuck, W. Poehlmann, Th. Pfeiffer, C. Koos, W. Freude, and J. Leuthold, "Cascade of 4 SOAs with 448 Gbit/s (224 Gbit/s) Dual Channel Dual Polarization 16QAM (QPSK) for High-Capacity Business Paths in Converged Metro-Access Networks," *Optical Fiber Communication Conference (OFC)*, Anaheim (CA), USA, March 2013, paper OTh4A.3, doi: 10.1364/OFC.2013.OTh4A.3. © 2013 The Optical Society.

Semiconductor optical amplifiers (SOA) for linear amplification of optical fields have gained interest as in-line amplifiers in optical communication networks [31, 130, 131],[C11]. For this purpose they are designed to have large saturation powers and a moderate gain. In this respect, these so-called linear SOA differ from nonlinear SOA, which are designed for small saturation powers and a large gain, and are used mainly for optical signal processing such as all-optical switching, regeneration, or wavelength conversion [105, 132, 133]. Some design guidelines for tailoring the saturation power of SOAs based on important physical parameters such as gain, material gain coefficient, differential gain, field confinement factor, carrier concentration, effective carrier lifetime, longitudinal SOA extension, and doping profile can be found in [134, 135].

Linear SOAs might become an attractive alternative to erbium-doped fiber amplifiers (EDFA), because SOAs are considered cost-efficient devices offering a broader gain spectrum (up to 100 nm) compared to EDFAs (typically 30 nm to 70 nm). This means that SOAs with fewer different center wavelengths are needed compared to EDFAs. In addition, the gain spectrum of SOAs can be centered at virtually any wavelength in a 10 THz window between 1250 nm and 1600 nm. Thus, SOAs are available for the O-band

(1300 nm), a range that could otherwise be covered by praseodymium-doped fiber amplifiers only, which are still relatively expensive. The possibility of integrating SOAs in photonic integrated circuits, their small form factor, and their small power consumption enable systems with high component density leading to lower installation and operational expenses.

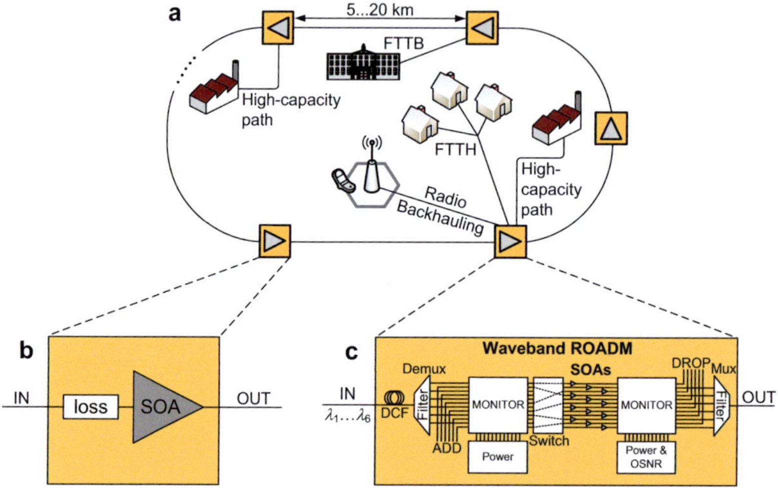

Fig. 3.1: Converged metro-access ring network scenario. (a) Infrastructures such as fiber-to-the-x (FTTx with x = B, H for building, home), radio backhauling, or high-capacity paths are supported. (b) SOAs along the ring compensate losses in the network so that a cascade of several SOAs results. (c) SOAs may be located in waveband ROADMs which provide data switching every 5 to 20 km. SOA = semiconductor optical amplifier, CWDM = coarse wavelength division multiplexing, ROADM = reconfigurable optical add-drop multiplexer, DCF = dispersion compensating fiber, Mux = multiplexer, Demux = demultiplexer, MONITOR = monitor tap, OSNR = optical signal-to-noise-ratio. Reprinted from [J1] © 2014 The Optical Society.

In the near future, optical metro networks and access networks will converge more and more, i.e., both the metro network infrastructure and the last mile infrastructure merge to one metro-access infrastructure [30, 131]. This converged metro-access network approach offers a large fiber-based capacity in a wide area. Fig. 3.1 depicts a converged metro-access ring network scenario (also see Fig. 1.7). With a feeder ring circumference of 30 to 60 km and a coverage area of about 300 km^2, this network serves a large number of subscribers (>100,000) in urban areas of mid-sized European cities. Such a network architecture offers virtually unlimited bandwidth for all kind of customers and infrastructures such as gigabit passive optical networks (GPON), fiber-to-the-x (FTTx with x = B, H for building, home), radio backhauling, and also high-capacity paths between enterprise customers, data centers, and large institutions such as universities or research centers. A high-capacity path from one enterprise to another offers, e.g., a 100 Gbit/s capacity which could be provided by the use of advanced modulation formats and/or polarization multiplexing such as 25 GBd DP-QPSK (dual polarization quadrature phase-shift keying) or 25 GBd 16QAM (quadrature amplitude modulation).

In the following, we first discuss the application of linear SOAs in a metro-access ring network, then explain an experimental setup with a linear SOA cascade for amplifying signals encoded with advanced modulation formats, and finally provide detailed experimental results.

3.1.1 Cascaded SOAs in Converged Metro-Access Networks

Previous SOA studies mainly focused on identifying the optimum operating conditions of a stand-alone SOA, and on increasing its linear operation range. These aspects and related issues have been extensively investigated for on-off-keying (OOK) data signals [136-143] as well as for PSK formats such as differential (D)PSK [144-148] and DQPSK [149, 150]. With the advent of coherent optical data communications, SOAs are required to tolerate signals encoded with advanced modulation formats such as QAM, which combines aspects of both amplitude-shift keying (ASK) and PSK formats. Some results have been already reported on the amplification of QAM signals by a stand-alone SOA [J5][151-153], and different schemes for pre-compensating [154] and post-compensating [155] SOA-induced nonlinearities in coherent QAM transmission systems have been proposed.

However, in many network scenarios a basic requirement is the capability of cascading several amplifiers. As an example, Fig. 3.1(a) depicts a ring network where SOAs are used to compensate the losses along the ring. Fig. 3.1(b) illustrates a simplified model comprising an SOA that compensates any kind of loss in the network. In a more detailed view as depicted in Fig. 3.1, the SOA may be part of a reconfigurable optical add-drop multiplexer (ROADM) that makes the network versatile and supports several wavelength bands (λ_1 to λ_6). The spectral width of a wavelength band may range from a few nanometers to a coarse wavelength division multiplexing (CWDM) 20 nm channel grid. Each band comprises several subchannels where data are transmitted employing different multiplexing techniques, modulation formats, and channel powers. For each band, one dedicated SOA compensates for the ROADM losses and for the transmission losses along the ring. Yet, in both cases a cascade of SOAs results.

Evidently, to support large data rates beyond 100 Gbit/s, the SOA cascade must cope with advanced modulation formats and multi-channel operation. Up to now, cascaded SOAs have only been examined for OOK [32, 33, 156] and for DPSK [144]. Virtually nothing is known about the practicality of an SOA cascade scenario with advanced modulation formats such as 8PSK as well as 16QAM and DP signals.

In this *Section*, we mimic a converged metro-access ring network scenario with a cascade of four commercially available linear SOAs. We investigate single λ-channel and dual λ-channel operation using single polarization (SP) and DP 28 GBd/channel binary (B)PSK, QPSK, 8PSK and 16QAM signals. This enables the transmission of aggregate data capacities of up to 448 Gbit/s. We evaluate the limiting factors, the achievable input power dynamic range (IPDR), polarization dependent issues, and the optimum SOA operating conditions. For the single λ-channel SP QPSK signal we also investigate the impact of different filter bandwidths (17 nm CWDM-filter, 2 nm, and no filter) after each SOA, as it influences on the achievable IPDR and the number of cascadable SOAs. To the best of our knowledge, this is the first detailed study of the performance of cascaded SOAs transmitting signals with advanced modulation formats.

The optimum SOA input power is found 11 dB below the 1 dB saturation input power, showing that linearity requirements for the SOAs are high and that the operating conditions of the network are primarily dictated by the SOA. When employing a 17 nm CWDM filter after each SOA, we observe a large IPDR of 20 dB for QPSK after the fourth SOA. For 8PSK, good IPDR values of 8…12 dB are obtained, and for 16QAM an IPDR of still 4…8 dB is achieved. For the single λ-channel SP QPSK signal, the use of a 2 nm filter bandwidth increases the IPDR by about 4 dB compared to the case with CWDM filter. However, without any filter after each SOA, the IPDR after the fourth SOA decreases by almost 10 dB.

3.1.2 Experimental Setup and SOAs under Test

Our experimental setup (Fig. 3.2) mimics a high-capacity path over up to four cascaded waveband ROADMs within the scenario of Fig. 3.1. It comprises a software-defined multi-format transmitter (Tx) [35] [83] encoding decorrelated data onto one or two optical carrier(s) at 1550.92 nm and 1549.72 nm (λ-channel spacing 150 GHz), and on one or two decorrelated polarization channels. The Tx symbol rate is 28 GBd. A coherent receiver

(Rx, Agilent N4391 optical modulation analyzer, OMA) processes the signals offline and provides dispersion compensation. For single λ-channel (dual polarization) operation, the bitrates are 28 Gbit/s (56 Gbit/s) BPSK (DP-BPSK), 56 Gbit/s (112 Gbit/s) QPSK (DP-QPSK), 84 Gbit/s (168 Gbit/s) 8PSK (DP-8PSK), and 112 Gbit/s (224 Gbit/s) 16QAM (DP-16QAM). For dual λ-channel (dual polarization) operation, the bitrates are 56 Gbit/s (112 Gbit/s) BPSK (DP-BPSK), 112 Gbit/s (224 Gbit/s) QPSK (DP-QPSK), 168 Gbit/s (336 Gbit/s) 8PSK (DP-8PSK), and 224 Gbit/s (448 Gbit/s) 16QAM (DP-16QAM).

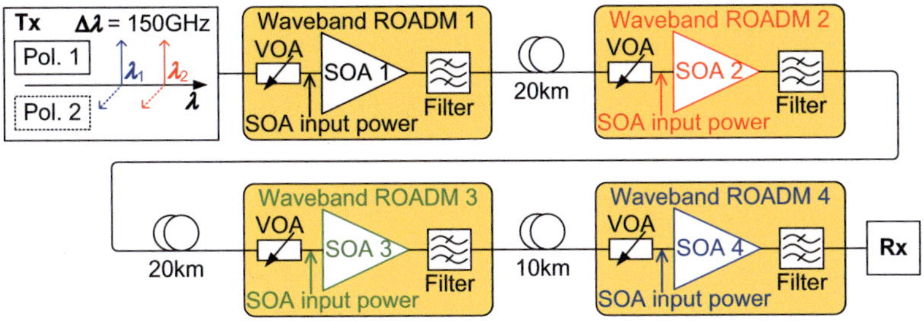

Fig. 3.2: Experimental setup with four cascaded waveband ROADMs to mimic a high-capacity path in a converged metro-access network (see Fig. 3.1). Identical SOAs in each waveband ROADM amplify the data signal which is then filtered. The total signal power into the first SOA is adjusted by the first VOA. All other VOAs are adjusted such that all SOAs have equal input power for a 0 dB net gain between subsequent SOA inputs. Reprinted from [J1] © 2014 The Optical Society.

All channels have equal average power. Each waveband ROADM is identically equipped with a variable optical attenuator (VOA) to mimic ROADM losses, an SOA under test, and a filter. The standard filter used is a 17 nm CWDM filter after each SOA. For the single λ-channel SP QPSK signal, we additionally investigate the two cases if a 2 nm filter or no filter is used. After the first, second and third waveband ROADM, the data signal is transmitted over 20 km, 20 km, and 10 km of standard single mode fiber (SMF).

We use four identical, commercially available SOAs (model CIP SOA-L-C-14-FCA) with 15 dB small signal fiber-to-fiber (FtF) gain (polarization dependency ±0.3 dB) at 1550 nm (bias current 500 mA), 6.5 dB noise figure (NF, polarization dependency ±0.3 dB), and a 1 dB saturation input power of −4 dBm. The representative FtF gain and NF for all four SOAs are shown in Fig. 3.3.

The total signal power into the first SOA is adjusted by the first VOA. All other VOAs are adjusted such that all SOAs have equal input power for a

0 dB net gain between subsequent SOA inputs. All SOAs have the same gain. The OMA input power is kept constant well above its sensitivity threshold for a bit error ratio (BER) of 10^{-9} for QPSK and 8PSK, and 10^{-6} for 16QAM. We analyze the error vector magnitude (EVM, see [J6] and *Appendix A.3.2*) after each ROADM. A detailed chart of the experimental setup is given in *Appendix A.6*.

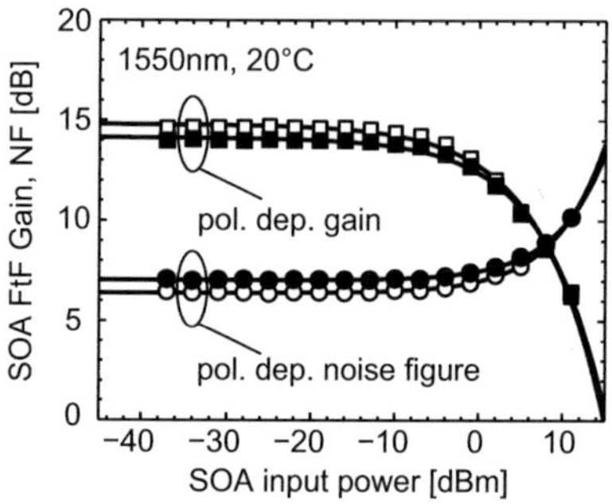

Fig. 3.3: Fiber-to-fiber (FtF) gain and noise figure (NF) of the SOAs used for the cascade experiment. All four SOAs have same characteristics. The 1 dB input saturation power is −4 dBm. Reprinted from [J1] © 2014 The Optical Society.

3.1.3 BER Estimation from EVM for Nonlinear Amplification

In our experiment, we are interested in the IPDR along the SOA cascade. We define the IPDR as the ratio of the upper and lower SOA input power limits, inside which the EVM corresponds to a BER better than 10^{-3}. For instance, the vertical dashed lines in Fig. 3.4(a) indicate the upper and lower SOA input power limits for a single SOA amplifying a single λ-channel SP-QPSK signal (56 Gbit/s). The horizontal line indicates a BER of 10^{-3}. At low input powers the signal quality is limited by amplified spontaneous emission (ASE) noise. At high input power levels, the signal quality is limited by SOA nonlinearities.

The EVM is a performance measure for advanced modulation formats. Under the assumption of additive white Gaussian noise (AWGN), BER values can be accurately calculated from measured EVM values [J6]. We measure EVM to judge the quality of our signal, because offline digital signal processing (DSP) and offline demodulation are used, which would make it very time consuming to reliably compute the BER. However, our definition of the IPDR is based on BER. Thus we first verify, for which SOA input power levels the assumption of AWGN is valid. We expect that the assumption does not hold in the nonlinear SOA regime.

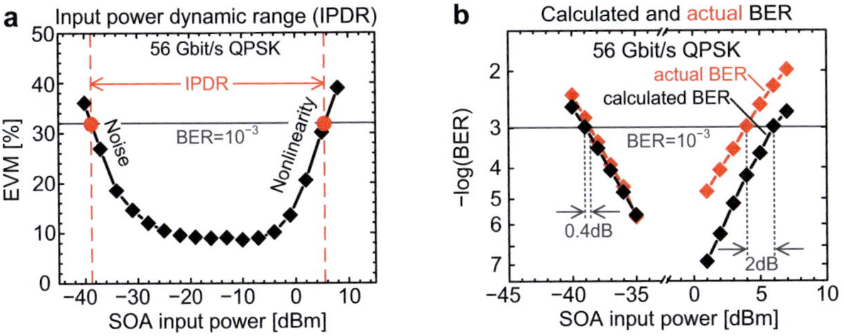

Fig. 3.4: (a) Error vector magnitude (EVM) as a function of the SOA input power and definition of the input power dynamic range (IPDR). The IPDR is defined as the ratio of the upper (nonlinearities) and lower (noise) SOA input power limits, inside which the EVM corresponds to a bit error ratio (BER) better than 10^{-3}. The upper and lower IPDR boundaries are indicated by the vertical dashed lines. (b) Relation between EVM and BER for low and large SOA input powers. Actual measured BER values and BER values calculated from EVM measurements assuming AWGN [J6] are compared for QPSK. Reprinted from [J1] © 2014 The Optical Society.

Fig. 3.4(b) compares actual measured BER values and BER values calculated from EVM measurements assuming AWGN for a single λ-channel SP-QPSK signal (56 Gbit/s). The signal was amplified by a single SOA. For a BER of 10^{-3}, the lower SOA input power limit obtained from actual measured BER values is only +0.4 dB larger than the lower input power limit obtained from the BER values calculated from EVM measurements, i.e., EVM is quite a reliable metric for the lower input power limit.

However, if the SOA amplifies nonlinearly, then the assumption of AWGN does not hold anymore. Then, the true upper power limit is smaller by −2 dB. In total, the true IPDR based on actual measured BER values is 2.4 dB smaller than the IPDR based on BER values which were calculated from EVM measurements. We did the same comparison for a single λ-channel SP-16QAM signal (112 Gbit/s). The deviation was +0.5 dB for the lower input power limit and −2.6 dB for the upper input power limit. To account for this uncertainty, we subsequently subtract 3 dB from all IPDR values obtained from EVM measurements.

3.1.4 SOA Cascade Limitations

In this *Subsection*, we discuss the EVM results obtained for the different modulation formats after each SOA along the cascade, if CWDM filtering is employed. For single λ-channel SP signals, we also analyze the phase and the magnitude error after each SOA. Furthermore, constellation diagrams are given to illustrate the limitations in the different operating regimes of the SOA. We also comment on the polarization dependence of the SOA cascade. From the measured EVM results in this *Subsection*, we determine the IPDR for each modulation format along the SOA cascade. The summary of the IPDR results is then given in *Subsection 3.1.5*.

SOA Cascade with BPSK Signals

In Fig. 3.5, the SOA cascade results for BPSK modulation are shown. Fig. 3.5(a) depicts the measured EVM after each of the four SOAs (SOA 1 black, SOA 2 red, SOA 3 green, SOA 4 blue) for single λ-channel (filled symbols) SP-BPSK as a function of the SOA input power. The solid horizontal line indicates the EVM limit corresponding to a BER of 10^{-3}. The dashed horizontal line indicates the back-to-back (BtB) signal quality without SOA cascade.

The associated constellation diagrams for the results given in Fig. 3.5(a) are shown in Fig. 3.5(g) for different SOA input power levels along the cascade. Additionally, we show in Fig. 3.5(b) the root-mean-square (rms) magnitude error (given as a percentage) and in Fig. 3.5(c) the rms phase error component (in degrees) of the EVM results of Fig. 3.5(a). The magnitude error is the difference in complex amplitude between the received symbol and the ideal constellation point normalized to the EVM normalization reference, which in our measurements is always the magnitude of the longest ideal constellation point. Knowing the different contributions of the magnitude error and the phase error to the overall EVM gives a more detailed insight by which mechanisms the signal is affected along the cascade.

At the best operating point with an input power of about −15 dBm (11 dB below the 1 dB gain saturation point) one finds a signal quality that is close to the BtB signal quality. At this operation point an SOA cascade with even more than four SOAs should work as well.

For low input powers, the performance is limited by amplified spontaneous emission (ASE) noise, leading to a symmetric broadening of the constellation points (see Fig. 3.5(g) for −28 dBm). From Fig. 3.5(b) and Fig. 3.5(c) it can be seen that in this region both the magnitude errors and the phase errors contribute roughly equally to the overall EVM.

For large SOA input power levels, crescent-shaped constellations can be observed after the first SOA (see Fig. 3.5(g), first row, + 5 dBm). Fig. 3.5(b) and Fig. 3.5(c) show that the dominating influence on EVM after the first SOA stems from phase errors. This is explained with zero-crossings of the field strength between symbols caused by the Tx used in this experiment [83]. The resulting gain change in the SOA then causes unwanted phase deviations by amplitude-phase coupling [J5] leading to a crescent-shaped constellation.

By cascading several SOAs, nonlinear (NL) distortions and ASE noise accumulate. When looking at the constellation diagrams in Fig. 3.5(g) for small input powers, one can observe how the noise and the symmetric broadening of the constellation diagrams increase for an increasing number of SOAs. In saturation (see −1 dBm), the distinct crescent-shaped constellation diagram is getting broader, the more SOAs are cascaded. This fact is also observed in Fig. 3.5(b), where the magnitude error for large input power levels is significantly increased if the signal is amplified by more than one SOA. This effect can be understood by the NF of the SOA (see Fig. 3.3), which grows in saturation. This increase in NF is particularly unfavorable when several SOAs are cascaded.

Fig. 3.5(d) depicts the measured EVM of one representative channel in the SP dual λ-channel operation (open symbols), where both λ-channels share the SOA gain. Since the total average SOA input power is depicted, the actual power per λ-channel is 3 dB lower, and hence the curves are shifted by about 3 dB to the right compared to single λ-channel operation, compare with Fig. 3.5(a). The same rules for the actual per-channel power apply also for the other modulation formats investigated in the following sections. The presence of a second λ-channel with a frequency spacing of 150 GHz to the first λ-channel shows no impairments. This is attributed to the almost continuous wave nature of phase-encoded signals. Thus one signal can be considered as a holding beam for the other signal [141]. The summary of IPDR values in Table 3.1 even indicates a slight increase in IPDR in the order of 1 dB for dual λ-channel BPSK and QPSK operation compared to single λ-channel operation.

Fig. 3.5(e) and Fig. 3.5(f) show the EVM for DP operation (dashed and dash-dotted lines) in one λ-channel (e), and for one representative λ-channel in a dual λ-channel operation (f). The horizontal dashed and dash-dotted lines indicate the BtB results for both polarizations. For dual λ-channel and DP operation, four independent signals are amplified, thus the input power per channel is reduced by 6 dB. Virtually no degradation in signal quality is seen for DP operation compared to SP operation. These results suggest that SOA cascades with considerably more than four SOAs are possible for DP-BPSK signaling.

SOA Cascade with QPSK and 8PSK Signals

In analogy to the BPSK results presented in Fig. 3.5, the EVM results and constellation diagrams for QPSK and 8PSK are shown in Fig. 3.6 and Fig. 3.7, respectively. In general, the same statements which were made for BPSK also hold for QPSK and 8PSK. All the three formats BPSK, QPSK, and 8PSK are phase-modulated signals with only one amplitude level, and are limited by the same effects inside the SOA cascade. However, for QPSK and 8PSK the distance between neighboring constellation points is smaller compared to BPSK. Thus, noise and/or nonlinearities will limit the performance already at higher/lower input power levels compared to BPSK. Consequently, the IPDR for both QPSK and 8PSK will be smaller than for BPSK.

In Fig. 3.6(g) and Fig. 3.7(g), the constellation diagrams for single λ-channel SP-QPSK and SP-8PSK, respectively, show only little impairments after four cascaded SOAs at an input power of −13 dBm which is close to the optimum operation point of −15 dBm. The good signal quality can also be seen in Fig. 3.6(a), Fig. 3.6(d), Fig. 3.7(a), and Fig. 3.7(d) showing the measured EVM along the SOA cascade for the single λ-channel (a) and dual λ-channel (d) SP signals. From these measurements, we conclude that single λ-channel and dual λ-channel SP-QPSK / 8PSK signals may also be transmitted over more than four cascaded SOAs.

For DP-QPSK, one can see from Fig. 3.6(e) and Fig. 3.6(f) that after the third SOA (green), the performance of one polarization is slightly worse than the performance of the second polarization and also worse than the performance of both polarizations after the fourth SOA (blue). We attribute this irregularity to the interplay between the random state of polarization in the transport fiber and the slightly polarization-dependent gain (PDG) of the SOA. The effect of the PDG on the SOA cascade will be discussed later. For DP-8PSK signals, Fig. 3.7(e) and Fig. 3.7(f), the irregularities due to polarization issues are more pronounced than for DP-QPSK.

SOA Cascade with 16QAM Signals

The results for the 16QAM signals are shown in Fig. 3.8. At small SOA input powers, the performance is again limited by ASE noise. In saturation, magnitude errors significantly con-tribute to the EVM already after amplification by the first SOA, see Fig. 3.8(b). The largest amplitude level (outermost symbols) is compressed and becomes (almost) indistinguishable from the middle amplitude level, see constellations diagrams for input power levels > -10 dBm in Fig. 3.8(g). Due to the proximity of the symbols in the constellation diagram, the IPDR will further decrease compared to QPSK and 8PSK. However, as can be seen from Fig. 3.8(a) and Fig. 3.8(d)–8(f), around the optimum SOA cascade input power, an SOA cascade comprising four SOAs is feasible. This enables high-capacity paths with data rates > 200 Gbit/s over 50 km SMF. However, irregularities due to polarization issues are seen for DP-16QAM in Fig. 3.8(e) and Fig. 3.8(f).

Optical Spectra along the SOA Cascade

The normalized spectra (resolution bandwidth RBW = 0.01 nm) after each SOA of the cascade for single λ-channel SP-QPSK and dual λ-channel SP-QPSK are shown in Fig. 3.9(a) and Fig. 3.9(b), respectively. The total SOA input power of the signals was $P_{\text{SOA}} = -13$ dBm (close to the optimum operation point of -15 dBm). Most of the ASE noise of the SOAs is filtered by a CWDM filter. It can be seen how the ASE noise accumulates along the SOA cascade within the CWDM filter bandwidth. The tilt of the ASE noise floor results from the wavelength dependent SOA gain.

Fig. 3.9(c) shows the normalized optical spectrum after the second SOA for dual λ-channel SP-QPSK for a total input power level of +2 dBm (non-linear regime) corresponding to a power of -1 dBm per λ-channel. Four-wave mixing (FWM) products can be seen 36 dB and 38 dB below the two channel peaks. These FWM products could deteriorate the signal quality in other λ-channels located at the same frequencies as the newly created FWM products. In practice, however, this is an unrealistic case, because the SOA cascade would not be operated at such large input powers where already a single λ-channel is severely distorted by amplitude-phase coupling as discussed before. The reason for signal degradation at low input powers, namely ASE noise accumulation, can be seen from the optical spectrum. However, for high input powers, the signal degrades primarily because of amplitude-phase coupling, and this cannot be seen from the spectrum. FWM products in the spectrum appear only at even larger input power levels.

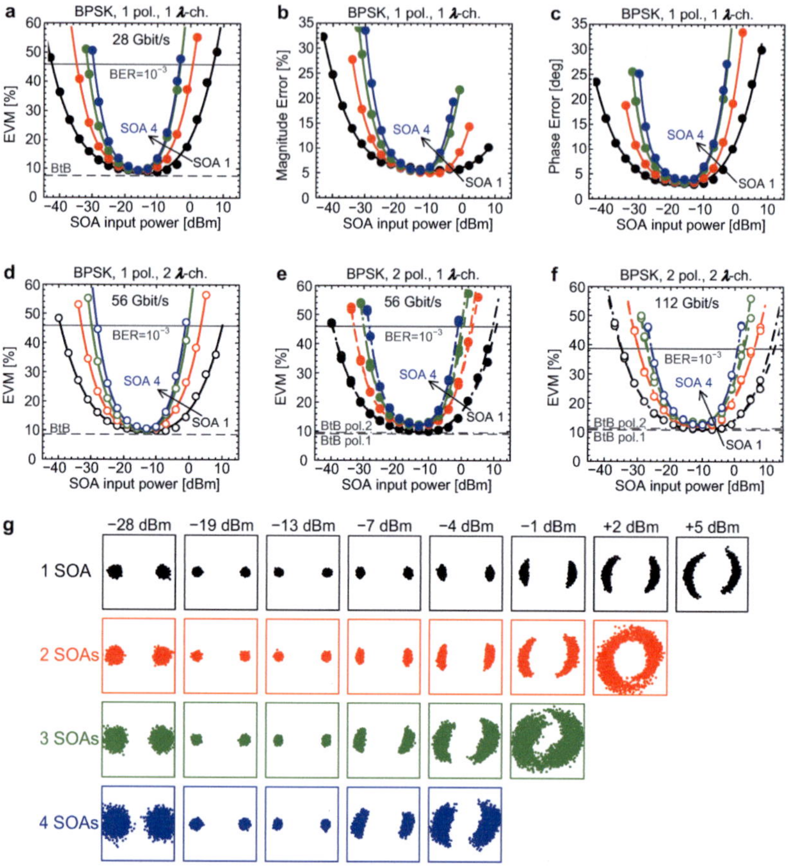

Fig. 3.5: EVM and constellation diagrams for BPSK signals in an SOA cascade. (a) EVM plot for the single λ-channel (filled symbols) single polarization (SP) signal, (b) magnitude error of the signal shown in (a), (c) phase error of the signal shown in (a). (g) Constellation diagrams of the single λ-channel SP constellation signal shown in (a) for different SOA input powers. (d) EVM plot of the dual λ-channel (open symbols) SP signal, (e) EVM plot of the single λ-channel dual polarization (DP, dashed and dash-dotted lines) signal, (f) EVM plot of the dual λ-channel DP signal. The solid horizontal lines indicate the EVM for a BER of 10^{-3}. Horizontal dashed/dash-dotted lines indicate the back-to-back (BtB) performance without SOAs. Reprinted from [J1] © 2014 The Optical Society.

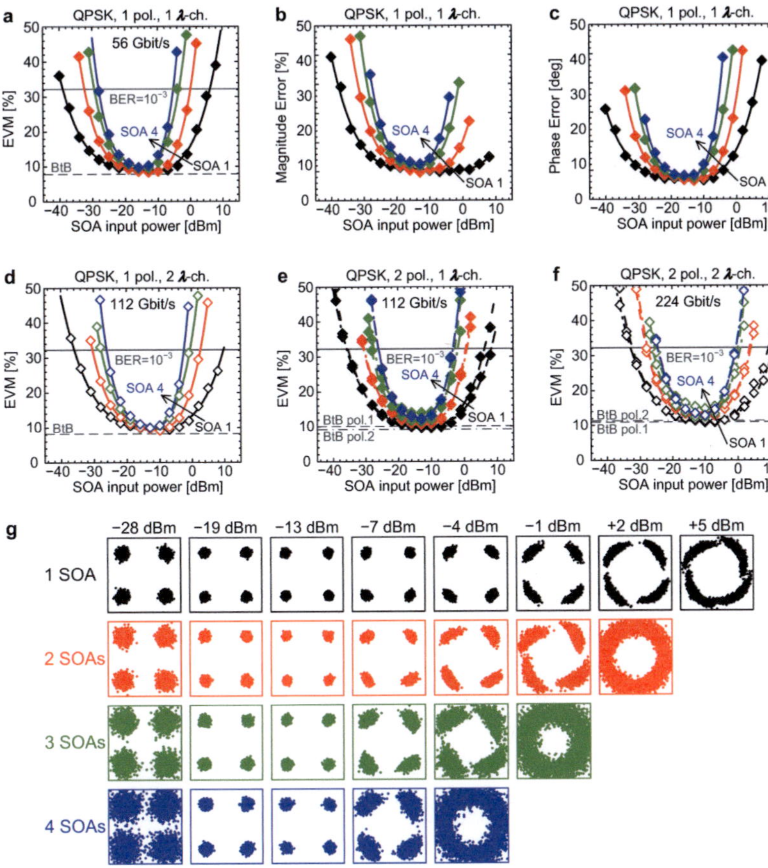

Fig. 3.6: EVM and constellation diagrams for QPSK signals in an SOA cascade. (a) EVM plot for the single λ-channel (filled symbols) single polarization (SP) signal, (b) magnitude error of the signal shown in (a), (c) phase error of the signal shown in (a). (g) Constellation diagrams of the single λ-channel SP constellation signal shown in (a) for different SOA input powers. (d) EVM plot of the dual λ-channel (open symbols) SP signal, (e) EVM plot of the single λ-channel dual polarization (DP, dashed and dash-dotted lines) signal, (f) EVM plot of the dual λ-channel DP signal. The solid horizontal lines indicate the EVM for a BER of 10^{-3}. Horizontal dashed/dash-dotted lines indicate the back-to-back (BtB) performance without SOAs. Reprinted from [J1] © 2014 The Optical Society.

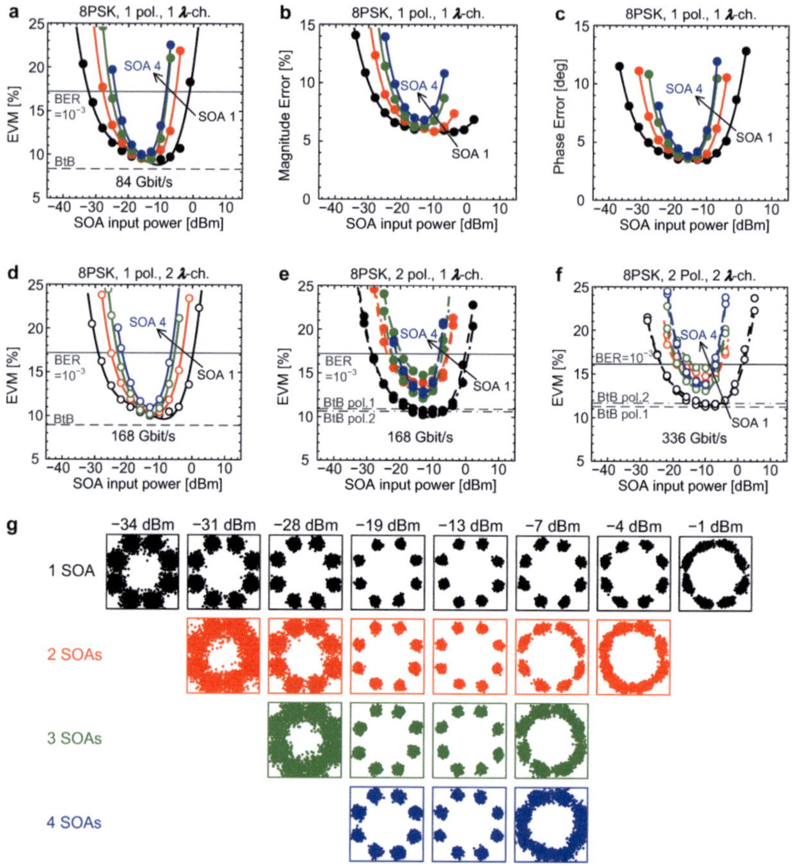

Fig. 3.7: EVM and constellation diagrams for 8PSK signals in an SOA cascade. (a) EVM plot for the single λ-channel (filled symbols) single polarization (SP) signal, (b) magnitude error of the signal shown in (a), (c) phase error of the signal shown in (a). (g) Constellation diagrams of the single λ-channel SP constellation signal shown in (a) for different SOA input powers. (d) EVM plot of the dual λ-channel (open symbols) SP signal, (e) EVM plot of the single λ-channel dual polarization (DP, dashed and dash-dotted lines) signal, (f) EVM plot of the dual λ-channel DP signal. The solid horizontal lines indicate the EVM for a BER of 10^{-3}. Horizontal dashed/dash-dotted lines indicate the back-to-back (BtB) performance without SOAs. Reprinted from [J1] © 2014 The Optical Society.

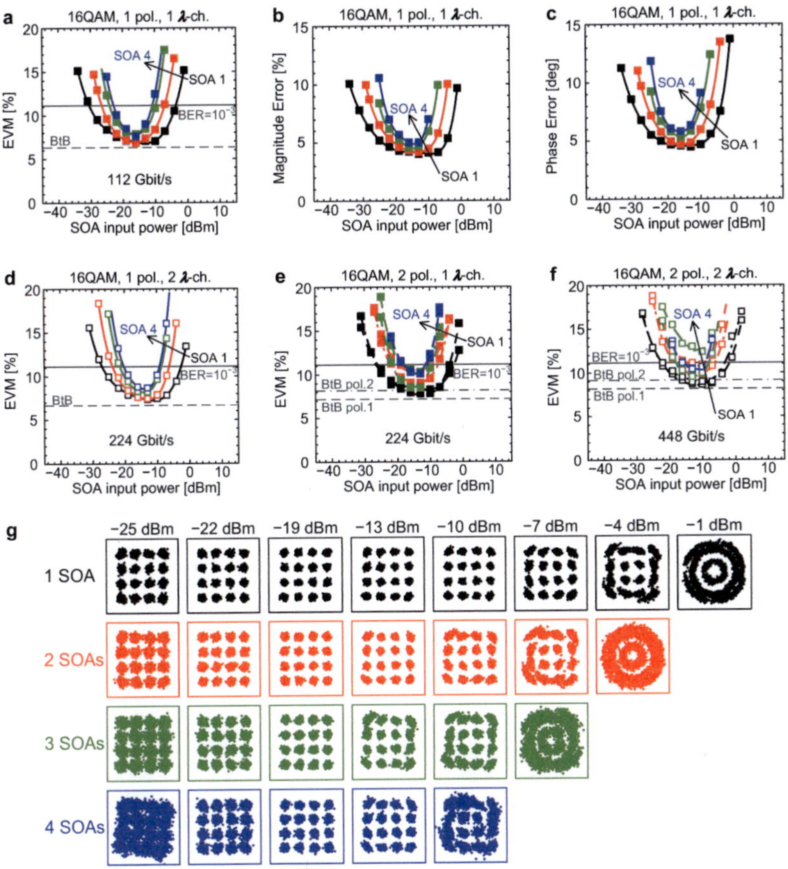

Fig. 3.8: EVM and constellation diagrams for 16QAM signals in an SOA cascade. (a) EVM plot for the single λ-channel (filled symbols) single polarization (SP) signal, (b) magnitude error of the signal shown in (a), (c) phase error of the signal shown in (a). (g) Constellation diagrams of the single λ-channel SP constellation signal shown in (a) for different SOA input powers. (d) EVM plot of the dual λ-channel (open symbols) SP signal, (e) EVM plot of the single λ-channel dual polarization (DP, dashed and dash-dotted lines) signal, (f) EVM plot of the dual λ-channel DP signal. The solid horizontal lines indicate the EVM for a BER of 10^{-3}. Horizontal dashed/dash-dotted lines indicate the back-to-back (BtB) performance without SOAs. Reprinted from [J1] © 2014 The Optical Society.

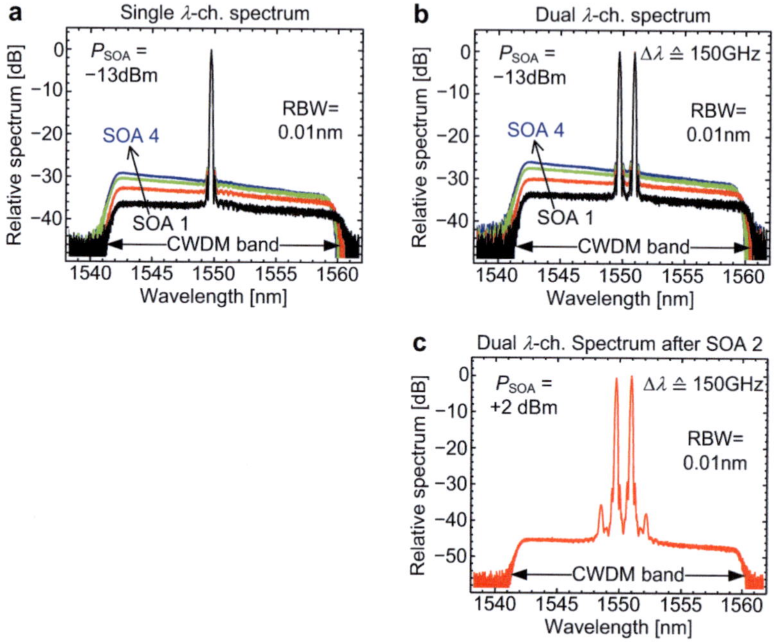

Fig. 3.9: Normalized optical spectra (resolution bandwidth RBW = 0.01 nm) along the SOA cascade for a total SOA input power of −13 dBm, (a) single λ-channel SP-QPSK, (b) dual λ-channel SP-QPSK. (c) Spectrum for dual λ-channel SP-QPSK signal after second SOA and a large total SOA input power of +2 dBm. Four-wave-mixing products can be seen. Reprinted from [J1] © 2014 The Optical Society.

SOA Cascade Polarization Dependence

An irregular performance of one polarization when amplifying DP signals, especially DP-16QAM, was observed and discussed previously. The irregularities are attributed to the interplay between the random state of polarization in the transport fiber and the slight PDG of the SOAs. By adjusting the input polarization at each SOA in the cascade to the minimum (maximum) gain, we evaluate the best (worst) case.

The effect of the PDG after one and four SOAs is depicted in Fig. 3.10(a) for a single λ-channel SP-QPSK signal and in Fig. 3.10(b) for a single λ-channel SP-16QAM signal. The upper power limit of the IPDR reduces by 2 to 3 dB, if the polarization is adjusted from minimum to maximum SOA gain.

In our SOA cascade experiment, we did not use any polarization controllers at the SOA because such an element would not be practical in a real network implementation. Our experiments suggest that the combination of random polarization in the transport fiber and a cascade of several SOAs with each SOA having a polarization dependency of "only" ±0.3 dB may lead to performance fluctuations and signal degradation. However, as we do not expect the PDG of SOAs to become much lower than ±0.3 dB, more detailed investigations on polarization as a limiting factor are required for quantifying this kind of degradation.

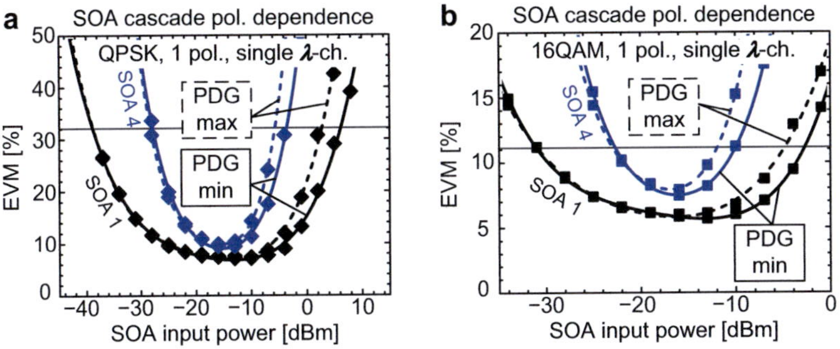

Fig. 3.10: Effect of polarization dependent gain (PDG) on SOA cascade. (a) single λ-channel SP-QPSK, (b) single λ-channel SP-16QAM. The upper power limit of the input power dynamic range (IPDR) reduces by 2 to 3 dB, if the polarization is adjusted from minimum to maximum SOA gain. Reprinted from [J1] © 2014 The Optical Society.

3.1.5 IPDR of the SOA Cascade with Advanced Modulation Formats

This *Section* discusses the IPDR results which were obtained from the EVM measurements in the previous *Section*. The IPDR values for the different modulation formats transmitted over the SOA cascade are then summarized in a table. We also discuss the influence of different filter bandwidths after each SOA on the IPDR along the cascade. For this purpose we transmit the single λ-channel SP-QPSK signal when employing a 2 nm filter and no filter after each SOA. The results are compared with those obtained for the single λ-channel SP-QPSK signal when employing a CWDM filter.

IPDR results for SOA cascade with CWDM filters

As described in *Subsection* 3.1.3, all IPDR values are corrected (reduced) by 3 dB to account for the difference between the actual BER and the BER values calculated from the EVM measurements in the nonlinear SOA regime where the assumption of AWGN does not hold.

Fig. 3.11(a) shows the IPDR as a function of the number of cascaded SOAs for SP signals. The IPDR results for DP signals are shown in Fig. 3.11(b) (polarization 1) and Fig. 3.11(c) (polarization 2), respectively. Single λ-channel results are represented by filled symbols, and dual λ-channel results are represented by open symbols.

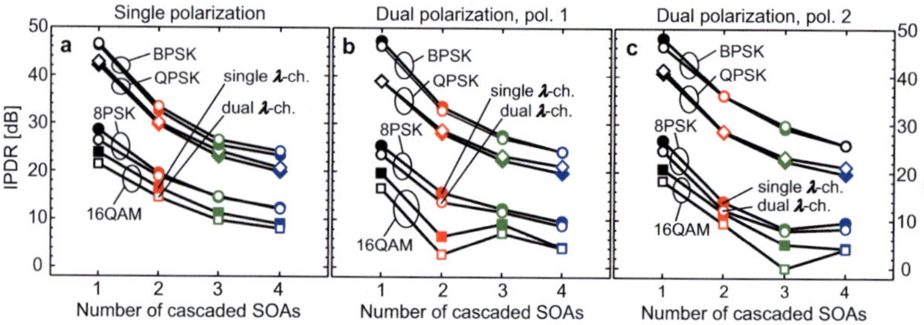

Fig. 3.11: Input power dynamic range (IPDR) as a function of the number of cascaded SOAs for (a) single polarization (SP) signals, (b) polarization 1 of dual polarization (DP) signals, (c) polarization 2 of DP signals. QPSK is a very robust signal. An IPDR of at least 20 dB is obtained after four SOAs. The DP-16QAM signals show some irregularities due to polarization dependent issues. Reprinted from [J1] © 2014 The Optical Society.

Both SP-QPSK and DP-QPSK formats exhibit an excellent IPDR of 20 dB after four SOAs, indicating that a high-capacity path with > 100 Gbit/s over a fiber length of 50 km is feasible. The QPSK signal is very robust and may also be employed in SOA cascades with more than four SOAs. This is sufficient for most of the application scenarios in future converged metro-access networks. Single-polarization 8PSK (SP-8PSK) signals offer a good IPDR of 12 dB (10 dB) after four cascaded SOAs. With SP-16QAM, an acceptable IPDR of 8 dB after four concatenated SOAs indicates a functional path. DP-16QAM still works with a small IPDR of 4 dB after four SOAs. However, due to polarization-related issues, irregularities after the second and third SOA can be seen.

Table 3.1 summarizes all IPDR values obtained in our measurements over a cascade of up to four SOAs and with CWDM filtering after each SOA. The QPSK signals are found to be very robust and represent a good trade-off between capacity and reach. QPSK offers twice the capacity of traditional OOK, and a cascade with more than four SOAs and QPSK modulation is feasible. The use of 16QAM offers capacities of nearly half a Tbit/s. However, especially DP-16QAM is only recommended over shorter distances.

Modulation			IPDR after n-th SOA [dB]				Data rate
Pol.	# of λ ch.	Format	$n=1$	$n=2$	$n=3$	$n=4$	[Gbit/s]
Single Polarization (SP)	1 λ ch	BPSK	46	32	25	23	28
	1 λ ch	QPSK	42	30	23	20	56
	1 λ ch	8PSK	29	19	15	12	84
	1 λ ch	16QAM	24	17	11	9	112
	2 λ ch	BPSK	47	33	27	24	56
	2 λ ch	QPSK	43	30	24	21	112
	2 λ ch	8PSK	26	19	15	12	168
	2 λ ch	16QAM	21	15	10	8	224
Dual Polarization (DP)	1 λ ch	BPSK	47	34	28	24	56
			46	33	27	24	
	1 λ ch	QPSK	39	28	22	20	112
			41	28	22	20	
	1 λ ch	8PSK	25	16	12	10	168
			27	14	8	10	
	1 λ ch	16QAM	20	7	9	4	224
			21	12	5	4	
	2 λ ch	BPSK	48	36	29	26	112
			46	36	30	26	
	2 λ ch	QPSK	39	29	23	21	224
			41	29	23	21	
	2 λ ch	8PSK	24	14	12	9	336
			25	12	8	8	
	2 λ ch	16QAM	17	3	7	4	448
			18	9	0	4	

IPDR color map: 0…5 dB | 6…19 dB | > 19 dB

Table 3.1: Summary of measured input power dynamic range (IPDR) values for a BER $< 10^{-3}$ along a cascade of four identical SOAs when using advanced modulation formats, a 0 dB net gain between subsequent SOA inputs, and CWDM filtering after each SOA. The two numbers given for DP signals denote the IPDR of both polarizations. The color grading (see legend below the Table) gives a hint to identify possible operation conditions in the network, [J1].

There is an IPDR trade-off between the modulation format that is used, which defines the capacity per channel, and the number of cascadable SOAs, which defines the distance that can be bridged in the ring network. In future access networks, resources can be allocated dynamically. For instance, a high-capacity path between two enterprises uses a single λ-channel with DP-QPSK modulation providing 112 Gbit/s capacity. On demand, this capacity can be increased to 224 Gbit/s either by adding another λ-channel with DP-QPSK modulation, or if a second λ-channel is not available, by changing the modulation format of the current λ-channel to DP-16QAM. Depending on the distance (number of ROADMs) between both enterprises, the IPDR could be insufficient for DP-16QAM. Then, a single λ-channel with DP-8PSK would be a possible trade-off providing a 168 Gbit/s capacity.

IPDR for SOA cascade with 2 nm filters and no filters

To investigate the influence of the filter bandwidth after each SOA on the IPDR, we replace the 17 nm CWDM filters in the setup of Fig. 3.2 by 2 nm filters. In a second case, we do not use any post-SOA filter at all. We transmit the single λ-channel SP-QPSK signal and compare the IPDR results to the results obtained with CWDM filtering.

In Fig. 3.12, the normalized spectra (resolution bandwidth RBW = 0.01nm) after each SOA of the cascade are shown for the cases when no filter is employed, Fig. 3.12(a), and or if a 2 nm filter is used, Fig. 3.12(b) (compare to Fig. 3.9(a) for CWDM filtering). When no filter is used, Fig. 3.12(a), all the ASE noise of one SOA is passed to the next SOA in the cascade. For low SOA input powers, the ASE noise becomes a major part of the total signal power. The subsequent SOA in the cascade will amplify the large amount of ASE from the previous amplifier and add its own ASE noise. By employing a filter after each SOA, the out-of-band ASE noise can be significantly reduced, see Fig. 3.12(b). For the same SOA input power, the ASE noise portion of the total signal power is reduced compared to the case when no filter is employed. It should be noted that for all three filtering cases, always the same Rx with a 0.6 nm Rx filter is used. Thus, the difference between the three considered cases is the portion of ASE noise along the SOA cascade with respect to the total signal power. The question is, how much does the IPDR along the SOA cascade change (improve) as a function of the filter bandwidth used after each SOA?

Fig. 3.13(a) depicts the upper and lower IPDR boundary (= upper and lower SOA input power limit) as a function of the number of cascaded SOA for the three different filter options (without filter, CWDM filter, 2 nm filter).

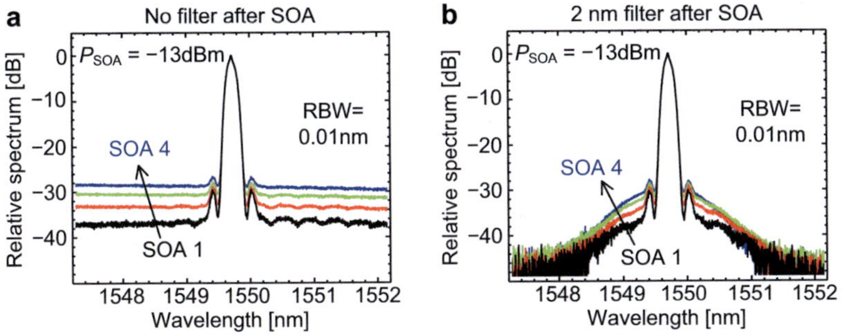

Fig. 3.12: Normalized optical spectra (resolution bandwidth RBW = 0.01 nm) for single λ-channel SP-QPSK along the SOA cascade. (a) Without filter after each SOA, (b) with a 2 nm filter after each SOA. The total SOA input power is −13 dBm. Reprinted from [J1] © 2014 The Optical Society.

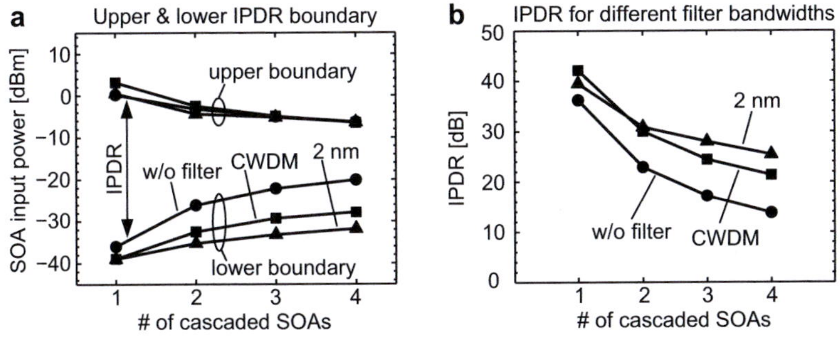

Fig. 3.13: Upper and lower boundary of the input power dynamic range (IPDR) as a function of the number of cascaded SOAs for different filter bandwidths (no filter, 17 nm CWDM filter, 2 nm filter). (b) IPDR as a function of the number of cascaded SOAs. By employing CWDM filters along the SOA cascade, the IPDR after the fourth SOA is increased by 7 dB compared to the case without filter. Reprinted from [J1] © 2014 The Optical Society.

It can be seen, that the upper IPDR boundary is virtually not affected by the filter bandwidth. The upper IPDR boundary is determined by nonlinear effects in the SOA. As expected, filtering has an influence on the lower IPDR boundary and thus on the achievable IPDR along the SOA cascade. For decreasing filter bandwidth (the measurement without filter represents the largest filter bandwidth) the lower IPDR boundary is shifting to lower SOA input powers, i.e. the IPDR is increasing. The quantitative effect of the different filter bandwidths on the IPDR along the SOA cascade is shown in Fig. 3.13(b). For the CWDM filter scheme, a 7 dB larger IPDR is achieved after the fourth SOA compared to the transmission without filter. For a filter bandwidth of 2 nm, the IPDR is 11 dB larger after the fourth SOA compared to the case without filter.

It should be pointed out that the use of a 2 nm filter would rather aim on filtering a single channel or an individual subchannel within a wavelength band, e.g., a CWDM wavelength band. However, this would considerably increase the waveband ROADM complexity. As demonstrated with our experiments, already the use of only CWDM filters considerably improves the IPDR along the SOA cascade. Table 3.2 summarizes the IPDR results for different filter bandwidths.

Single λ-channel SP-QPSK Filter after each SOA	IPDR after n-th SOA [dB] $n = 1$	 $n = 2$	 $n = 3$	 $n = 4$	Data rate [Gbit/s]
No filter	36	23	17	14	56
17 nm filter (CWDM)	42	30	24	21	56
2 nm filter	39	31	28	25	56

IPDR color map: 0…5 dB | 6…19 dB | > 19 dB

Table 3.2 Influence of different filter bandwidths on the input power dynamic range (IPDR) along a cascade of four identical SOAs with single λ-channel SP-QPSK and a 0 dB net gain between subsequent SOA inputs. The use of a CWDM filter gives a 7 dB larger IPDR after four SOAs compared to the case if no filter is used.

3.1.6 Conclusions

We demonstrate that signals encoded with advanced modulation formats such as BPSK, QPSK, 8PSK, and 16QAM can be amplified by at least four cascaded SOAs. This enables high-capacity paths with a high data capacity in the order of a Tbit/s for future converged metro-access ring networks.

A large IPDR of > 20 dB is found for SP-QPSK and DP-QPSK signals after four cascaded SOAs, indicating that even a larger number of SOAs can be cascaded. With SP-8PSK and DP-8PSK, an acceptable IPDR of about 10 dB after four cascaded is achieved. Also the demanding 16QAM signals can be transmitted over four SOAs, after all with an IPDR of 4…8 dB. Thus, the robust QPSK signal is recommended for large transmission distances. More complex formats are a viable option for very large capacities over shorter distances.

Despite the small PDG of ±0.3 dB of the SOAs used in this paper, our experiments show irregular performance of one polarization when amplifying DP signals, especially for DP-16QAM. This is attributed to the interplay between the random state of polarization in the transport fiber and the slightly polarization-dependent SOA gain. We recommend the usage of SOAs having the smallest possible polarization dependence of the gain.

The best SOA performance is found at input powers 11 dB below the 1 dB input saturation power, indicating that only linear SOAs can be used. For low input powers, the performance is limited by the accumulation of ASE noise along the SOA cascade. Thus, SOAs with a low noise figure are desirable to enlarge the achievable IPDR at the low input power side.

For phase-modulated signals (BPSK, QPSK, and 8PSK) the limitations after the first SOA (if operated in the nonlinear region) mainly result from amplitude-phase coupling. Gain changes during zero-crossing field strength transitions induce phase errors. For 16QAM, large amplitude changes are less likely. However, the SOA gain compression leads to the compression of the higher amplitude levels (magnitude errors). After more than one SOA, also the phase modulated signals suffer more and more from magnitude errors for large input power levels. The additional magnitude errors were attributed to the increasing NF of the SOA in saturation, and to the effect of cascading several SOAs which are all operated in their nonlinear regime. For an increasing number of SOAs, the upper and lower input power limit come closer and thus the IPDR decreases.

The dual λ-channel transmission experiments reveal only little signal quality impairments compared to the single λ-channel transmission experiments. We believe that due to the relatively weak amplitude variation for the investigated advanced phase modulation formats the amplification of such signals is almost comparable to the amplification of continuous wave signals. Both channels are amplified independently, and degradations due to inter-channel cross-talk (cross-gain and cross-phase modulation) are weak. This is in contrast to the amplification of multi-channel OOK signals, where inter-channel crosstalk is a major issue when operating the SOA in saturation.

Semiconductor optical amplifiers offer a large gain bandwidth. Filtering after each SOA in an SOA cascade is beneficial to reduce unwanted out-of-band noise. We show that already the use of a CWDM filter (17 nm bandwidth) after each SOA enhances the IPDR by 7 dB after the fourth SOA compared to the case when no filter is used. When using a 2 nm filter bandwidth after each SOA, the IPDR after the fourth SOA increases by another 4 dB compared to the CWDM filter case.

3.2 SOAs for Radio-over-Fiber Signals

In *Section 3.1*, (cascaded) SOAs were investigated with advanced modulation formats. In the context of optical-wireless convergence (*Section 1.2*), also radio-over-fiber (RoF) signals will be transmitted and distributed in future access networks. Therefore it is necessary to study the influence of the SOA gain dynamics on RoF signals. Besides the IPDR, which we call optical power dynamic range (OPDR) for RoF signals, we are also interested in the electrical power dynamic range (EPDR). Furthermore, we investigate the small signal optical modulation response of an SOA in the linear and nonlinear region for intensity modulated optical signals carrying sinusoidal RF signals with a frequency of up to 20 GHz.

3.2.1 Frequency Response Measurement of an SOA

In *Section 2.4.3* and *Appendix A.1*, the optical small signal response of an SOA was derived in theory. In this *Subsection*, we experimentally determine the frequency response of a quantum-dot (QD) and a commercially available quantum-well (QW) SOA. Both devices will be used in various RoF transmission experiments in the following *Sections*.

Characteristics of the SOAs under Test

Two different types of SOAs with similar unsaturated gain values are investigated. The first SOA is a QD SOA. It has a gain peak wavelength of about 1295 nm at an operating temperature of 20 °C and a 3 dB gain bandwidth of about 35 nm. The FtF gain for different operating currents at a wavelength of 1300 nm is shown in Fig. 3.14(a). The QD SOA gain has a polarization dependency of 10 dB due to the nature of QD. For an operating current of 300 mA, the unsaturated FtF gain is 16.7 dB and the 1 dB and 3 dB saturation input powers are −7 dBm and −1 dBm, respectively. The noise figure of the QD SOA is 10 dB. The QD SOA was fabricated at the Technical University of Berlin (group of Prof. Dr. Bimberg). Further information on the QD SOA structure and fabrication process is given in [157].

The second SOA under test is a commercially available QW device provided by Alphion (SAO20i, receiver pre-amplifier/ in-line amplifier). The gain peak wavelength is at 1278 nm and the SOA has a 3 dB gain bandwidth of 60 nm. The measured FtF gain for different operating currents is shown in Fig. 3.14(b). The maximum polarization dependency of the gain is specified to be 1.2 dB. For a current of 300 mA, the unsaturated FtF gain is 18 dB and the 1 dB and 3 dB saturation input powers of the QW SOA are –10 dBm and –4 dBm, respectively. The noise figure of the QW SOA is 6 dB.

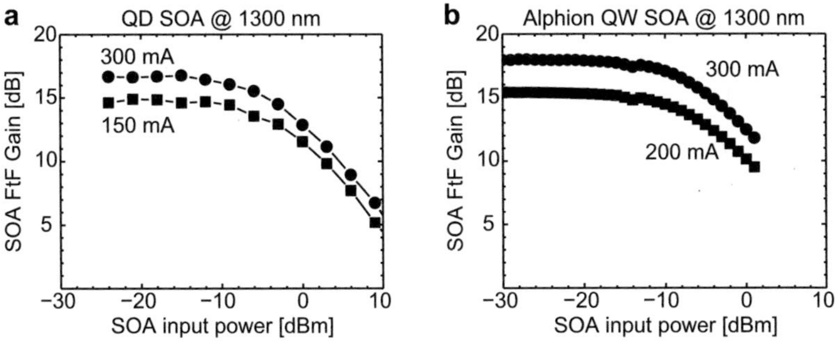

Fig. 3.14: Fiber-to-fiber (FtF) gain at 1300 nm as a function of the SOA input power for the tested SOAs. (a) Quantum dot (QD) SOA and (b) commercial quantum well (QW) SOA (Alphion SAO20i).

Experimental Setup

The frequency response of the SOAs is measured using a lightwave component analyzer (LCA). The measurement setup is depicted in Fig. 3.15. The LCA comprises a RF network analyzer and an optical transmitter and receiver. To measure the small signal frequency response of an optical device, a sinusoidal RF signal is first encoded on a continuous wave laser by optical intensity modulation with a Mach-Zehnder modulator (MZM) which is operated at its quadrature point. The modulation depth is about 10 %. The modulated signal passes through the optical device under test and is received at the LCA with a photodetector (square law detection). The RF output signal from the photodetector is then fed back to the network analyzer unit, where the signal's magnitude and phase are compared to the original RF signal which was modulated on the optical carrier at the LCA transmitter. The employed LCA can measure the frequency response in the range of 0.13…20 GHz.

In our setup, we boost the optical output power from the LCA with a praseodymium-doped fiber amplifier (PDFA) to reach also very large SOA input powers. The PDFA does not change the modulation depth of the optical signal. A 2 nm bandpass filter after the PDFA reduces the ASE noise added by the PDFA. The average SOA input power P_{in} is adjusted with an optical attenuator. The small signal variation of the SOA input power P_{in} due to the RF modulation of the optical carrier is denoted as ΔP_{in}. We use a polarization controller in front of the SOA because the QD SOA has a strongly polarization dependent gain. After the SOA, a second 2 nm filter reduces the ASE noise added by SOA.

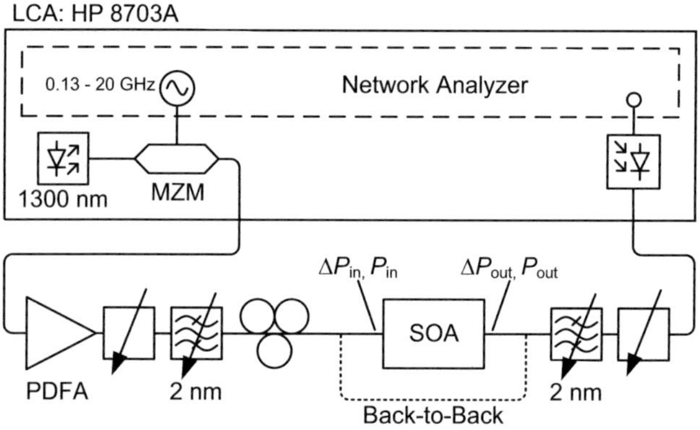

Fig. 3.15: Measurement setup to determine the frequency response of an SOA. A lightwave component analyzer (LCA) modulates the intensity of an optical carrier with a sinusoidal RF signal. The signal passes through the SOA under test and is received at the LCA with a photodetector. The magnitude and phase of the received RF signal is compared to the original RF signal in the network analyzer. To obtain the frequency response, this procedure is repeated for different RF frequencies in a frequency range of 0.13…20 GHz. A praseodymium-doped fiber amplifier (PDFA) in front of the SOA is used to reach also large SOA input power levels and investigate the frequency response in the nonlinear SOA regime. The back-to-back measurement without SOA is to calibrate the system.

The system was calibrated without SOA (back-to-back measurement with fiber patch cord) for a certain power level at the LCA's photodetector. The same measurement conditions at the LCA's photodetector must be guaranteed irrespective of the SOA output power P_{out}. Therefore, an optical attenuator is used after the SOA to adjust the input power to the LCA's photodetector to always the same value as used in the calibration measurement. However, this means that the input power to the LCA is reduced by the actual SOA gain G (which depends on P_{in}) and the modulation response H_{m} is measured where m_{out} and m_{in} denote the optical modulation depth at the SOA output and input, respectively, and $H(f)$ is the small-signal frequency response as introduced in *Section 2.4.3*:

$$H_{\text{m}}(f)=\frac{1}{G}H(f)=\frac{1}{G}\frac{\Delta P_{\text{out}}(f)}{\Delta P_{\text{in}}(f)}=\frac{P_{\text{in}}}{P_{\text{out}}}\frac{\Delta P_{\text{out}}(f)}{\Delta P_{\text{in}}(f)}=\frac{m_{\text{out}}(f)}{m_{\text{in}}(f)}\ . \tag{3.1}$$

Measurement Results

Fig. 3.16 shows the measured (red) and fitted (black) magnitude and phase of the modulation response $H_m(f)$ for different SOA input powers of both the QD SOA, Fig. 3.16(a,b) and the QW SOA, Fig. 3.16(b,c). Both SOAs show qualitatively the same behavior. The linear and nonlinear SOA operation regions show different modulation responses. The fitted curves are based on Eq. (2.34) to (2.36). From the fit curves, the SOA time constants τ_1, τ_2, and the effective carrier lifetime τ_c are determined in *Subsection 3.2.2*.

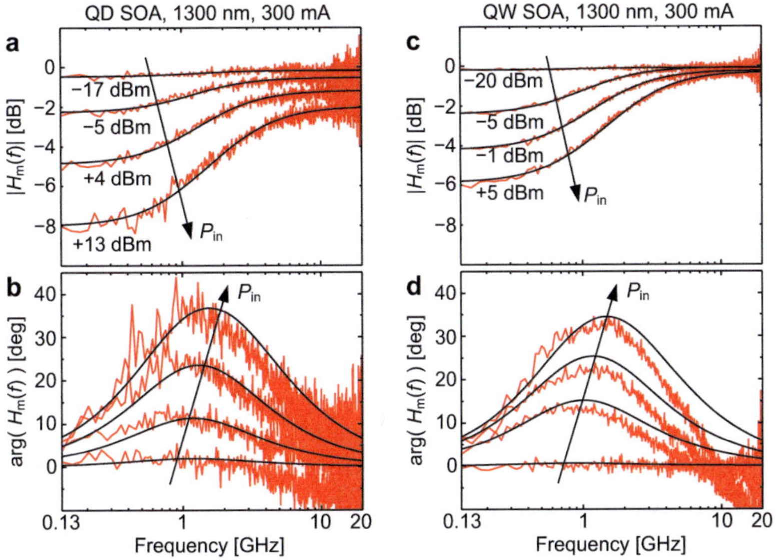

Fig. 3.16: Measured (red lines) and fitted (black lines) magnitude and phase of the modulation response $H_m(f)$ of the investigated QD SOA (a,b) and QW SOA (c,d) for different SOA input powers. In the linear SOA operation region, a nearly flat response is observed, while in the nonlinear SOA regime, the modulation response strongly depends on the modulation frequency.

In the linear regime (low input powers of −17 dBm and −20 dBm in Fig. 3.16), a virtually flat magnitude and phase response is observed. In the linear operation region the SOA does not influence the RF signal which is modulated on the optical carrier.

However, with increasing SOA input power the modulation response changes. In the nonlinear regime (see curves for +13 dBm and +5 dBm in Fig. 3.16), the modulation response has a high-pass characteristic. As can be seen from the magnitude response of the investigated QW SOA, Fig. 3.16(c), slow RF modulation signals with frequencies $f < 10$ GHz are attenuated, while signals with $f > 10$ GHz are virtually not affected. For the QD SOA, the same behavior can be seen for slow RF modulation signals. However, for fast RF modulation signals with $f > 10$ GHz, the magnitude of the modulation response does not reach 0 dB within the frequency measurement range. The SOA gain dynamics which lead to the observed frequency response will be discussed in the next *Subsection*.

Note that Fig. 3.16(a,c) show the magnitude of the *optical* modulation response as measured in our setup with the LCA. The magnitude of the *electrical* frequency response can be calculated by squaring the magnitude of the optical modulation response. For instance, the magnitude of the optical modulation response for the QD SOA ($P_{in} = +13$ dBm, at very slow modulation frequencies) is attenuated by 8 dB which actually means an attenuation by 16 dB of the electrical frequency response.

3.2.2 Impact of Gain Saturation Dynamics on RoF Signals

In the last *Subsection*, different modulation responses were observed depending on the SOA operation region. Here, the influence of the SOA gain dynamics on the magnitude of the optical modulation response is discussed.

In the linear SOA regime, i.e., at average input powers P_{in} much smaller than the saturation input power, the average SOA output power P_{out} is linearly proportional to P_{in}, Fig. 3.17(a). The RF modulation of the optical intensity by the LCA leads to a variation ΔP_{in} of the SOA input power P_{in}. The constant unsaturated SOA gain $G_0 = P_{out}/P_{in}$ relates the input power variation ΔP_{in} to the output power variation $\Delta P_{out} = G_0 \Delta P_{in}$. Thus, for a linear operation of the SOA, the optical modulation depth m_{out} at the SOA output equals the modulation depth m_{in} at the SOA input irrespective of the RF modulation frequency,

$$m_{in} = \frac{\Delta P_{in}}{P_{in}} = \frac{\Delta P_{out}}{P_{out}} = m_{out} , \qquad (3.2)$$

which results in a constant modulation response $H_m(f) = 0$ dB.

In the nonlinear SOA regime, two cases must be distinguished. For slow modulation speeds (for $f \to 0$), Fig. 3.17 (b) left, the SOA dynamics can follow the sinusoidal fluctuation of the optical intensity leading to a compression of the output signal due to the nonlinear relationship between P_{out} and P_{in}. Thus the optical modulation depth at the SOA output m_{out} is smaller than the modulation depth m_{in} at the SOA input, $m_{out} < m_{in}$. The stronger the

SOA is operated in saturation, the smaller becomes the magnitude of the modulation response $H_m(f)$.

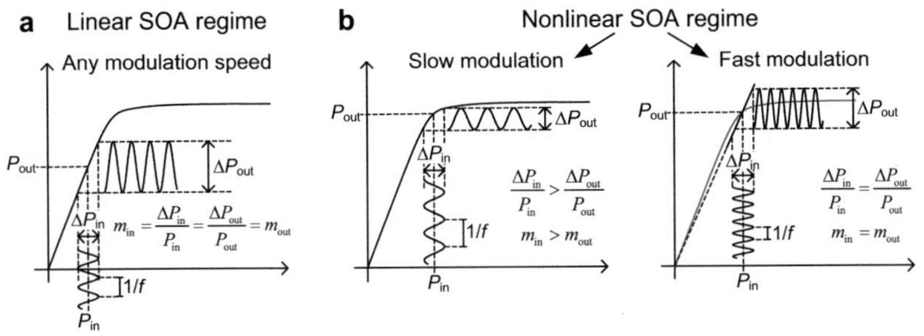

Fig. 3.17: SOA modulation response explained for (a) the linear and (b) the nonlinear SOA regime. In the linear SOA regime, the average SOA output power P_{out} is linearly proportional to P_{in}. The RF small signal modulation of the optical intensity causes a variation ΔP_{in} of the SOA input power P_{in}. The constant unsaturated SOA gain G_0 relates the input power variation ΔP_{in} to the output power variation $\Delta P_{out} = G_0 \Delta P_{in}$. The optical modulation depth m_{out} at the SOA output equals the modulation depth m_{in} at the SOA input irrespective of the modulation frequency. In the nonlinear SOA regime, two cases must be distinguished. For slow modulation speeds, the SOA dynamics is able to follow optical power variation leading to a compression of the output signal due to the nonlinear relationship between P_{out} and P_{in}, thus $m_{out} < m_{in}$. For fast modulation speeds, the SOA dynamics cannot follow the input power variations. A quasi-linear operation with an average gain $G = (P_{out}/P_{in}) < G_0$ leads and m_{out} approaches m_{in}.

For fast modulation speeds ($f \rightarrow \infty$), Fig. 3.17 (b) right, the SOA dynamics cannot follow the input power variations. This results in a quasi-linear operation, where the signal is amplified with an average gain $G = P_{out} / P_{in}$ smaller than the unsaturated SOA gain G_0. However, the modulation depths m_{out} approaches m_{in} and $H_m(f) \rightarrow 0$ dB and thus a quasi-linear amplification of the RoF signal is achieved in the nonlinear SOA regime. In our measurement for the QD SOA, Fig. 3.16(a), the magnitude of the modulation response does not reach 0 dB at the upper measurement limit (20 GHz) and a residual attenuation is still visible compared to the measurement results for the QW SOA in Fig. 3.16(c). We assume that the modulation response of the QD SOA is determined by more than one time constant, which are not included in our small signal model given in *Section 2.4.3*. However, we expect that the response reaches 0 dB for even larger frequencies because of by the fast carrier refilling (in the order of few ps) of QD states by the reservoir states surrounding the dots [104].

Estimation of the Effective Carrier Lifetime

The magnitude of the modulation response $|H_{\mathrm{m}}(f)|$ can be used to determine the effective carrier lifetime τ_{c}. For this purpose, the measurements results (red lines) from Fig. 3.16(a,c) are fitted (black lines) using a slightly modified version of the modulation response $H_{\mathrm{m}}(f)$ from Eq. (3.1). A real valued amplitude adjustment factor A is introduced for better fit results:

$$|H_{\mathrm{m}}(f)| = \left| A \frac{\tau_2}{\tau_1} \frac{1 + \mathrm{j}2\pi f \tau_1}{1 + \mathrm{j}2\pi f \tau_2} \right| . \qquad (3.3)$$

The fit yields the parameters τ_1, τ_2, and A. As shown in *Section 2.4.3*, the effective carrier lifetime can be approximated by $\tau_{\mathrm{c}} \approx \tau_1$ if the condition $(G_0 - 2) >> 2\ln(2) P_{\mathrm{in},0} / P_{\mathrm{sat}}^{\mathrm{in}}$ holds. For an input power of P_{in} = +5 dBm for both investigated SOAs, the left hand side of this condition is about 6 dB larger than the right hand side. For smaller input powers than that, the left hand side becomes even larger.

Table 3.3 gives the effective carrier lifetime values $\tau_{\mathrm{c}} \approx \tau_1$ in picoseconds and the amplitude adjustment factor A as determined from the fitted modulation response for both SOAs at an operating current of 300 mA and for different input power levels. The determined effective carrier lifetime is in the order of 200 ps.

(a) QD SOA, I = 300 mA

P_{in} [dBm]	τ_{c} [ps]	A [lin]
−17	139	0.96
−5	163	0.88
+4	186	0.76

(b) QW SOA, I = 300 mA

P_{in} [dBm]	τ_{c} [ps]	A [lin]
−20	188	0.98
−5	205	0.99
−1	213	0.95
+5	204	0.94

Table 3.3: Effective carrier lifetime τ_{c} for different SOA input powers P_{in} determined from the fitted modulation response measurement for (a) the QD SOA and (b) the QW SOA. For an amplitude adjustment factor $A = 1$, the fit model corresponds to the modulation response given in Eq. (3.1).

For both SOAs, the effective carrier lifetime τ_{c} increases with increasing input power. At small input powers, the spontaneous recombination $1/\tau_{\mathrm{c}}$ rate dominates, while at large input power the carriers primarily recombine due to stimulated transitions and the spontaneous recombination rate decreases, $1/\tau_{\mathrm{c}}$ becomes smaller. For the QW SOA, Table 3.3(b), the amplitude adjustment factor A is very close to 1 irrespective of the SOA operating region, showing that our small-signal model fits very well for this SOA. As mentioned before, we assume that an additional time constant is needed to accurately describe the small-signal response of the QD SOA for large input powers. There, the amplitude adjustment factor for the QD SOA clearly deviates from 1, Table 3.3(a).

3.2.3 Optical and Electrical Power Dynamic Range

The content of this Subsection has been published in the conference publication [C25]:

> **S. Koenig**, J. Pfeifle, R. Bonk, T. Vallaitis, C. Meuer, D. Bimberg, C. Koos, W. Freude, and J. Leuthold, "Optical and electrical power dynamic range of semiconductor optical amplifiers in radio-over-fiber networks," *36th European Conference and Exhibition on Optical Communication (ECOC'10)*, Torino, Italy, September 2010, paper Th.10.B.6, doi: 10.1109/ECOC.2010.5621535.
> Reprinted, with permission, from [C25] © 2010 IEEE.

Radio-over-Fiber (RoF) systems are used for the transport and distribution of radiofrequency (RF) signals such as used in wireless local area networks (WLAN), WiMAX, or mobile communications over a fiber infrastructure [23]. Currently, the convergence of gigabit passive optical networks (GPON) and RoF systems is discussed, see *Section 1.2*. For reach extended GPON, SOAs have emerged as cost effective and viable amplifiers. However, with respect to RoF signals, this leads to the question of the SOA's optical power dynamic range (OPDR), as customers are located at different distances to the SOA, and whether a large electrical power dynamic range (EPDR) can be maintained. Actually, a large EPDR is desirable in the face of RF path fading (exceeding 30 dB), movement of costumers inside radio cells, and RF-to-optical conversion inefficiencies. Up to now, a study of OPDR and EPDR for SOA in such RoF networks is still missing.

In this *Subsection*, we show that SOA (we use the QD SOA from *Section 3.2.1*) provides a huge 30 dB optical and 30 dB electrical power dynamic range. EPDR can be optimized at the expense of OPDR and vice versa. Experiments have been performed with QPSK and 16-QAM RoF signals at 10 MBd.

Experimental RoF Setup

To determine the OPDR and the EPDR, the QD SOA is placed between the transmitter (Tx) and the receiver (Rx) of the experimental setup as shown in Fig. 3.18. The transmitter comprises of a 1.3 µm cw laser (output power P'_0, center frequency f_{opt}) which is fed into a MZM biased at its quadrature point. Digitally modulated QPSK and 16QAM signals at a symbol rate of 10 MBd are generated by a vector signal generator (VSG) at RF carrier frequencies of $f_{RF} = 2$ GHz and 5 GHz. The RF signals with power P_E from the VSG are fed into the MZM. The RF signals reside as sidebands on the intensity modulated optical carrier. The optical signal is then amplified by a PDFA and the optical power levels are adjusted by a variable optical attenuator (VOA) in front of the SOA. For the remainder of this *Subsection*, P_O denotes the power of the optical carrier from the Tx into the SOA. Moreover, m denotes the optical modulation depth of the optical carrier induced by the electrical power P_E of the RF signal modulated onto this optical carrier.

The optical power into the pre-amplifier receiver (Rx) is kept constant at −15 dBm in order to maintain the same sensitivity. The RoF signals are demodulated by a vector signal analyzer (VSA) and the root-mean-square (RMS) error vector magnitude (EVM) normalized to the average signal power is measured. An EVM of 16.7 % for QPSK and 7.5 % for 16QAM, respectively, is calculated to correspond to a bit error ratio (BER) of 10^{-9} (error-free operation) [158]. The minimum required P_E for error-free operation without the SOA is −32 dBm (2 GHz) and −30.5 dBm (5 GHz) for QPSK and −24 dBm (2 GHz) and −23.5 dBm (5 GHz) for 16QAM, respectively.

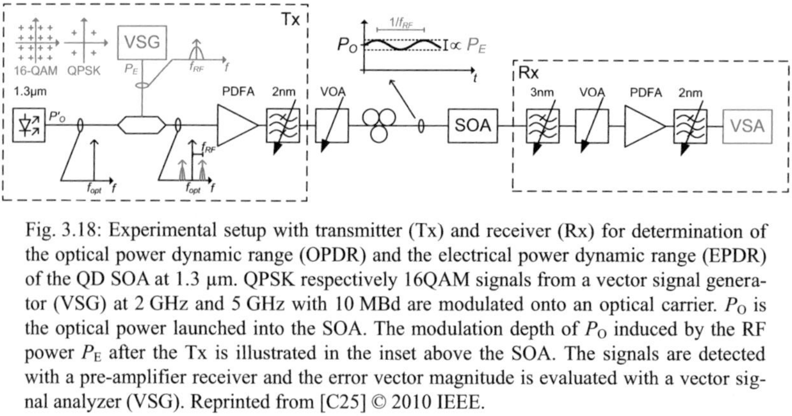

Fig. 3.18: Experimental setup with transmitter (Tx) and receiver (Rx) for determination of the optical power dynamic range (OPDR) and the electrical power dynamic range (EPDR) of the QD SOA at 1.3 µm. QPSK respectively 16QAM signals from a vector signal generator (VSG) at 2 GHz and 5 GHz with 10 MBd are modulated onto an optical carrier. P_O is the optical power launched into the SOA. The modulation depth of P_O induced by the RF power P_E after the Tx is illustrated in the inset above the SOA. The signals are detected with a pre-amplifier receiver and the error vector magnitude is evaluated with a vector signal analyzer (VSG). Reprinted from [C25] © 2010 IEEE.

EVM Measurements and RF Power Penalty

EVM measurements are carried out for both QPSK and 16QAM signals at 2 GHz and 5 GHz. The measurement range for P_E and for P_O are

−30 dBm < P_E < +5 dBm (corresponding to an optical modulation depth of 0.7 % < m <37 %) and −24 dBm < P_O < +15 dBm, respectively.

Fig. 3.19(a) shows a map of EVMs as a function of P_O for different P_E for the 16QAM signal on the 5 GHz carrier. The dashed line indicates the EVM limit for error-free operation. At low P_O, the signal performance degrades as the device is operated below the optical signal-to-noise ratio (OSNR) limit. At high P_O, the device is operated in gain saturation leading to a higher EVM. For subsequent use, the OPDR is defined as the power range within which error-free amplification is guaranteed for a particular P_E. For example, in Fig. 2(a) the lower ($P_{1,O}$) and upper ($P_{2,O}$) limits provide an OPDR > 30 dB for P_E > −17.5 dBm (m > 2.8 %).

Fig. 3.19(b) shows a map of EVMs as a function of P_E for different P_O. The EVM decreases for increasing P_E. Best performance is achieved for P_O close to the SOA saturation input power. The EPDR is defined in analogy to the OPDR as the RF power range within error-free operation is obtained for a given P_O.

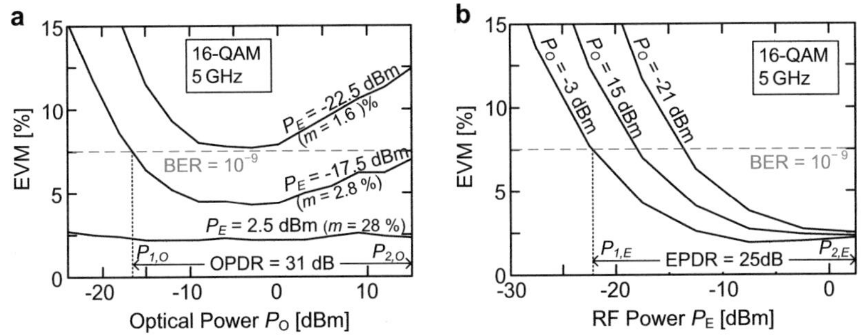

Fig. 3.19: Map of EVM for measured combinations of P_O and P_E operation points. (a) EVM as a function of optical power into QD SOA. The larger the RF power level P_E, the larger the OPDR. (b) EVM as a function of RF power P_E. The largest EPDR exceeding 30 dB is offered if the optical input power P_O is near the SOAs input saturation power, which here is at P_O = −3 dBm. Reprinted from [C25] © 2010 IEEE.

While the SOA provides gain, i.e., budget extension, it also induces signal degradation. This degradation may be overcome by modulating the RF signal at a higher power. This additional power is defined as the RF power penalty. It is plotted in Fig. 3.20(a) for QPSK and 16QAM at both 2 GHz and 5 GHz. It is observed to be independent of the modulation format. The minimum RF power penalty of 1 dB for the 5 GHz signals and 2 dB for the 2 GHz signals occurs at P_O = −5 dBm. At lower P_O, the signal performance is degraded by noise and the power penalty increases independently of the carrier frequency.

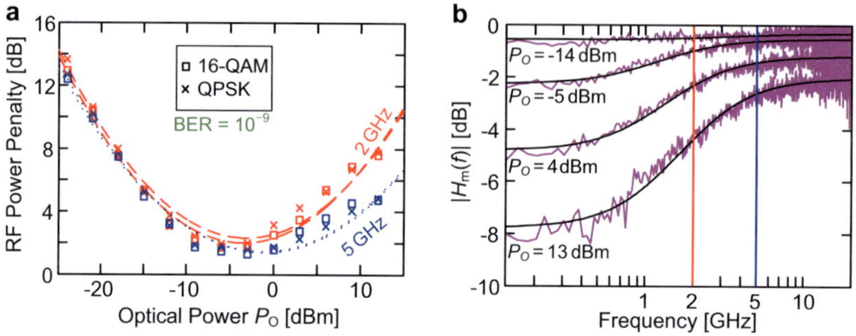

Fig. 3.20: (a) Additional RF power needed in order to maintain the 10^{-9} BER level in the system with an SOA. The RF power required is plotted as a function of the optical input power levels for QPSK and 16QAM at 2 GHz and 5 GHz. At low optical power, the RF power penalty increases as the device is operated below the optical signal-to-noise ratio limit. At high optical power, the device is operated under gain saturation which also increases the penalty. (b) Modulation response of QD SOA for different power levels showing the origin of the RF carrier frequency related penalty at higher optical input power levels P_O. Reprinted from [C25] © 2010 IEEE.

However, at high P_O the degradation is stronger for the 2 GHz signals than for 5 GHz signals. This is mostly due to the frequency dependence of the gain in saturation. This can be deduced from the magnitude of the modulation response $|H_m(f)|$ of the SOA, see *Section 3.2.1*. Fig. 3.20(b) shows the magnitude of the modulation response from 0.1 GHz to 20 GHz for different P_O at 1.3 µm. For low P_O, the RF response is flat, whereas for high P_O the RF response has high-pass characteristic. Under deep gain saturation and for low RF frequencies (0.1 GHz to 1 GHz), the SOA gain can follow the small signal modulation. Therefore, the low power levels of the optical envelope are amplified more than the high power levels. This decreases the optical modulation depth and results in a reduced RF response. At higher RF frequencies, the SOA small signal gain cannot follow (slow time constant (~200 ps) [104]) leaving the modulation depth unchanged, see *Section 3.2.2*.

OPDR and EPDR for System Applications

Fig. 3.21(a) shows the OPDR with its lower ($P_{1,O}$) and its upper ($P_{2,O}$) limits over P_E. An ultra-large OPDR > 40 dB is observed for sufficiently high P_E. With an OPDR > 40 dB, the QD SOA works well independently of the distance to the customer, i.e., with a weak or a strong RoF signal. For weak P_E the ideal OPDR range is small and the optical power P_O into the SOA must be chosen reasonably large. As a consequence it should be used as an inline amplifier in the upstream of an extended GPON, e.g., with 1:32 split ratios.

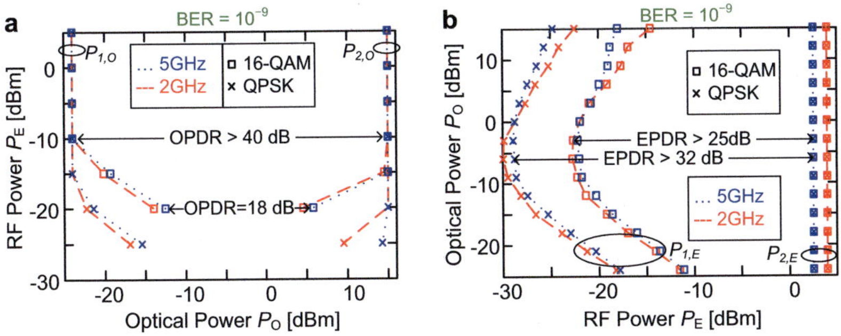

Fig. 3.21: Map of (a) OPDR operation ranges for various P_E and (b) EPDR operation ranges for various P_O for the considered RoF signals. In (a) an ultra-large OPDR > 40 dB is observed for RF powers > −15 dBm (m > 3.7 %) for QPSK and > −10 dBm (m > 6.7 %) for 16QAM. (b) Lower and upper limit of the EPDR for all optical power levels. Maximum EPDR > 32 dB are observed for QPSK and EPDR > 25 dB for the more demanding 16QAM at –5 dBm optical power. Reprinted from [C25] © 2010 IEEE.

Fig. 3.21(b) shows the lower ($P_{1,E}$) and the upper ($P_{2,E}$) limits for P_E and thus the EPDR for error-free operation for all measured P_O. Maximum EPDR > 32 dB for QPSK and EPDR > 25 dB for the more demanding 16QAM are observed at the ideal P_O. Thus, in scenarios where a large EPDR is needed, the SOA should be operated around $P_O = -5$ dBm.

3.2.4 Conclusions

Due to SOA gain saturation dynamics, the optical modulation response of an SOA operated in saturation has a high-pass characteristic and impacts RoF signals depending on the modulation frequency. For slow modulation speeds, the SOA dynamics can follow the RoF signal leading to a compression of the amplified signal. For fast modulation speeds, the SOA dynamics cannot follow the input power variations which results in a quasi-linear amplification of the RoF signal. The effective carrier lifetime τ_c can be extracted from the measured SOA modulation response.

SOAs are well suited to extend the reach of RoF PON networks. Experiments have been performed with 2 GHz and 5 GHz RoF signals and 1.3 µm SOA. We showed that SOAs can be operated within a large OPDR (> 40 dB) and EPDR (≈ 30 dB). The OPDR and the EPDR are interrelated. One quantity can be optimized at the expense of the other: On the one hand, for a large OPDR, a large modulation depth is needed which in turn limits the EPDR. On the other hand, for a large EPDR, the SOA needs to be operated around the saturation input power, which limits the available OPDR.

3.3 SOA Reach-Extender in a WDM-TDM-GPON Converged with RoF

Parts of this Section were published in the Conference publication [C23]:
S. Koenig, M. Hoh, R. Bonk, H. Wang, P. Pahl, T. Zwick, C. Koos, W. Freude, and J. Leuthold, "Rival signals in SOA reach-extended WDM-TDM-GPON converged with RoF," *Optical Fiber Communication Conference (OFC)*, Los Angeles (CA), USA, March 2011, paper OWT1, doi: 10.1364/OFC.2011.OWT1.

As discussed in *Section 1.2* and *Section* 1.3, radio-over-Fiber (RoF) systems are tentatively to converge with gigabit passive optical networks (GPON) [22]. Of particular interest are reach extended GPON, where SOAs have emerged as cost effective and viable amplifiers.

In *Section 3.2.3*, we showed that SOA exhibit a large optical power dynamic range (OPDR) and electrical power dynamic range (EPDR) and therefore fulfill the basic needs of extended PON networks. In next generation converged RoF and GPON, however, the RoF signal and the GPON signals are amplified simultaneously by the same reach extender SOA. The question thus is how the RoF signal is influenced by a rival signal (e.g., a WDM signal, a bursty TDM-GPON signal, or another RoF signal), as both signals are amplified at the same time by the same SOA.

In this *Section*, we study the SOA as a reach extender and the impact of a rival WDM signal, a rival TDM-PON, and a rival RoF signal on a RoF subscriber in a converged radio-over-fiber and WDM–TDM–GPON upstream scenario (1.3 µm band). We use the Alphion QW SOA which was introduced in *Section 3.2.1*. It is shown that the SOA can handle rival signals well – if the network operator takes appropriate measures. More precisely, the optical powers of the network subscribers must not exceed the saturation limit of the extended reach PON SOA.

3.3.1 Network Scenario and Experimental Setup

Fig. 3.22 shows a converged RoF and WDM–TDM–GPON upstream scenario (1.3 µm band) with a SOA as reach extender between the central office (CO) and the WDM multiplexer. The scenario has been set up to study the impact of a rival WDM-PON channel (continuous NRZ-OOK at λ_{WDM}), the impact of a rival TDM-GPON-like upstream burst signal (λ_{TDM}), and the impact of a rival RoF signal (RoF2 at λ_{RoF2}) on the investigated RoF signal

(RoF1 at λ_{RoF1}), when the rival signal and the RoF1 signal are amplified simultaneously by the SOA. We investigate the impact of only one type of rival signal at a time on the RoF1 signal. The rival signals operate at a wavelength of $\lambda_{\text{WDM}} = \lambda_{\text{TDM}} = \lambda_{\text{RoF2}} = 1310$ nm, while the investigated RoF1 uses the wavelength $\lambda_{\text{RoF1}} = 1300$ nm.

For the RoF1 signal and the different types of rival signals shown in Fig. 3.22 we study the impact of several network related parameters on the RoF1 signal quality:

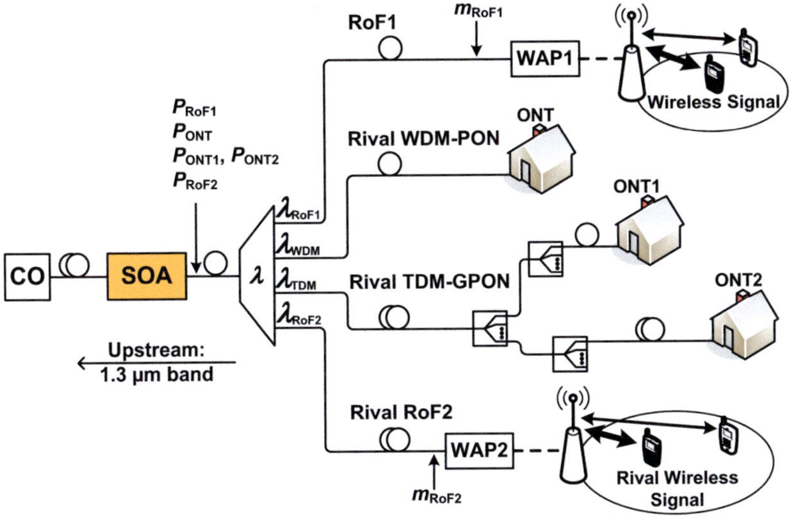

Fig. 3.22: Converged RoF and WDM–TDM–GPON upstream scenario (1.3 µm band) with a SOA as reach extender between the CO and the WDM multiplexer. The two branches RoF1 (λ_{RoF1}) and RoF2 (λ_{RoF2}) lead to wireless access points (WAP). The TDM–GPON (λ_{TDM}) includes passive splitters with different split ratios. We consider the bursty upstream signal of two optical network terminations (ONT) with burst length 100 µs. The impact of a rival WDM channel (λ_{WDM}) with continuous NRZ-OOK on the RoF1 signal is studied for different average optical input power P_{ONT} (this corresponds to different distances of the ONT to the SOA). Impact of a bursty 1.25 Gbit/s NRZ-OOK TDM-GPON data signal: Different distances of the TDM-GPON to the SOA and/or different split ratios are emulated by varying the average optical input power P_{ONT1} from ONT1 into the SOA. Different burst ratios (BR) emulate displacements of ONT2 with respect to ONT1.The impact of a rival RoF signal RoF2 onto RoF1 is studied for different average optical input powers P_{RoF1} and P_{RoF2} into the SOA (this corresponds to different distances of the WAP to the SOA) and different optical modulation depths m_{RoF1} and m_{RoF2} (this corresponds to different RF power levels at the WAP originating from different mobile users).

First we study the performance of the RoF1 signal λ_{RoF1} without the influence of any rival signal. We vary the average optical power P_{RoF1} of the RoF1 signal into the SOA to emulate different distances of a wireless access point 1 (WAP) to the reach extending SOA. In addition, we vary the optical modulation depth m_{RoF1} to emulate different RF power levels at the WAP originating from different mobile users who have different distances to the WAP1.

Second, we study the impact of the rival WDM-PON channel λ_{WDM} on the RoF1 signal λ_{RoF1}. We vary the average optical power P_{ONT1} at the SOA input to emulate an optical network termination (ONT) that has different distances to the SOA.

Third, the impact of a bursty TDM-PON-like signal λ_{TDM} is studied against the average optical input power P_{ONT1} into the SOA and the burst ratio *BR* (see *Section 1.1.2*) Different P_{ONT1} represent different distances of the TDM-GPON to the SOA and/or different split ratios of the passive splitters, whereas different burst ratios *BR* correspond to displacements of ONT2 with respect to ONT1.

Last, the impact of the rival RoF2 signal λ_{RoF2} onto the RoF1 signal λ_{RoF1} is studied when varying the average optical power P_{RoF1} and P_{RoF2}. This corresponds to a scenario where RoF WAPs of subscribers 1 and 2 are located at different distances to the SOA. In addition we vary the optical modulation depths m_{RoF1} and m_{RoF2} to emulate different RF power levels at the WAP originating from different mobile users who have different distances to the WAP.

Experimental Setup

Fig. 3.23 shows a simplified schematic of the experimental implementation of (a) the RoF1 transmitter (Tx), (b-d) the rival Tx, and (e) the RoF receiver (Rx). Refer to *Appendix A.6* for a more detailed setup chart.

Radio-over-fiber Transmitters

The two RoF signals RoF1 and RoF2 in our experiment are generated by two vector signal generators (VSG), Fig. 3.23(a,d). The RoF1 signal is a Gray-coded QPSK signal at a symbol rate of 20 MBd at a carrier frequency of f_{RoF1} = 5GHz. The rival RoF2 signal is a 16QAM signal with the same symbol rate and same carrier frequency of f_{RoF2} = 5GHz. At this carrier frequency, the RoF signals are not much affected by the modulation response when the SOA is operated in the nonlinear regime, see *Section3.2.1*. The VSGs also perform filtering of the RF signals with a raised root cosine filter having a roll of factor of 0.35.

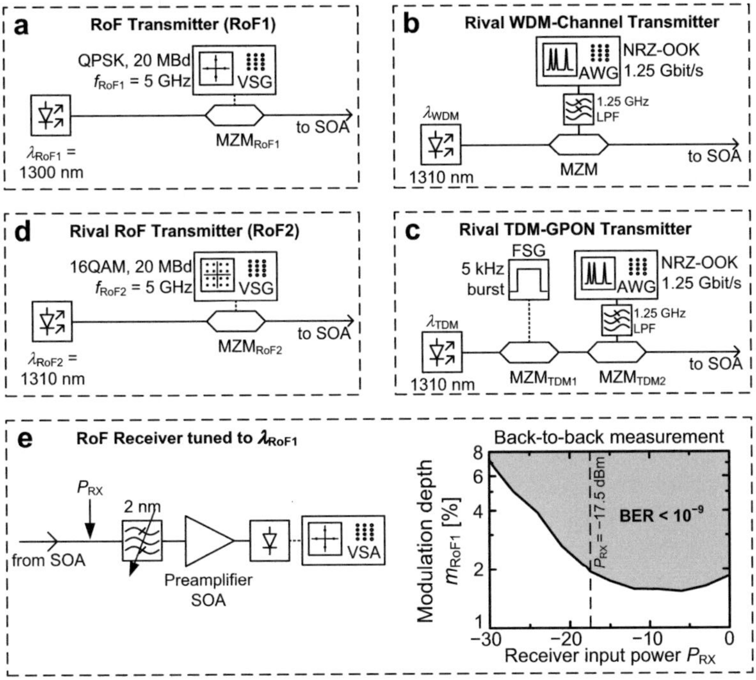

Fig. 3.23: Simplified experimental implementation of (a) the RoF1 transmitter (Tx), (b) the rival WDM-channel Tx, (c) the rival TDM-GPON transmitter, (d) the rival RoF2 Tx, and (e) the RoF receiver (Rx) with back-to-back measurement (inset) without the SOA under test. The RoF1 signal and one rival signal are simultaneously amplified by the SOA (not drawn). The RoF Rx is tuned to the wavelength of the RoF1 signal to study the impact of the rival signal on RoF1. AWG = arbitrary waveform generator, FSG = function signal generator, LPF = low pass filter, MZM = Mach-Zehnder modulator, VSA = vector signal analyzer, VSG = vector signal generator.

A WAP is represented by a continuous wave (cw) laser (λ_{RoF1} = 1300 nm, λ_{RoF2} = 1310 nm) which is externally modulated by a Mach-Zehnder modulator (MZM) biased at its quadrature point (intensity modulation). The RF signals from both VSG are fed into the MZM. We define the average modulation depth m as the ratio between the peak-to-peak RF voltage amplitude and the V_π-voltage of the MZM. We determine the RF peak-to-peak voltage from the average RF power provided by the VSG assuming a sinusoidal signal and a 50 Ω impedance. In our experiment, the average optical modulation depths $m_{RoF,n}$ may be varied between $1.2\,\% < m_{RoF1} < 37.5\,\%$ and $1.4\,\% < m_{RoF2} < 14.4\,\%$, respectively.

To reach also the nonlinear SOA regime, the RoF signals are optically amplified after modulation using a PDFA (RoF1 Tx) and a booster SOA (operated in its linear regime, RoF2 Tx), not drawn in Fig. 3.23(a,d). The optical powers P_{RoF1} and P_{RoF2} to the SOA under test are then adjusted by a variable optical attenuator (VOA) in each RoF branch.

Rival WDM-Channel Transmitter

In the rival WDM-channel Tx, Fig. 3.23(b), a continuous 1.25 Gbit/s NRZ-OOK signal is modulated on the optical carrier at a wavelength of $\lambda_{\text{WDM}} = 1310$ nm using a MZM biased at its quadrature point (intensity modulation). The NRZ-OOK data signal is generated by an arbitrary waveform generator (AWG). We use a pseudo random bit sequence (PRBS) with length 2^7-1. After the AWG, the data signal is filtered with an electrical low-pass filter with a cut-off frequency of 1.25 GHz. Thus the electrical spectra of the rival NRZ-OOK signal and the RoF1 signal do not overlap. The optical signal is amplified by a booster SOA (operated in its linear regime) and the optical input power P_{ONT} to the SOA under test is adjusted by a VOA, not shown in Fig. 3.23(b).

Rival TDM-GPON Transmitter

The rival TDM-GPON Tx, Fig. 3.23(c), generates a 1.25 Gbit/s (PRBS 2^7-1) upstream signal that emulates the burst traffic from two ONTs. The TDM-GPON Tx is a modified version of the WDM-channel Tx. The optical burst signal is generated with the help of a function signal generator (FSG) and a MZM_{TDM1} prior to the data modulator MZM_{TDM2}. The FSG outputs a rectangular shaped signal at a frequency of 5 kHz which results in a rectangular modulation of the optical power of the cw carrier λ_{TDM}. The bias point of MZM_{TDM1} is always adjusted such, that one level of the rectangular signal leads to maximum power transmission. By varying the output amplitude of the FSG, the burst ratio *BR* between the two ONTs can be changed. The burst length of both ONTs is 100 μs (rise/fall time measured to 20 ns) and *BR* may be varied between 0 dB (continuous NRZ-OOK signal) and 10 dB. The 1.25 Gbit/s NRZ-OOK data from an AWG are encoded onto the rectangular optical envelope using the data modulator MZM_{TDM2}. A VOA after MZM_{TDM2} is employed to adjust the optical input power to the investigated SOA.

Fig. 3.24 shows the output signal of the burst mode transmitter for different *BR* measured with a digital communications analyzer (DCA). When the DCA is triggered on the burst frequency of 5 kHz, Fig. 3.24(a-c), then the 100 μs time slots of ONT1 and ONT2 become visible. The greenish area between the one and zero level indicates the transitions between the data bits at a data rate of 1.25 Gbit/s. Horizontal dashed lines in Fig. 3.24(b) indicate the optical power levels for a one level of ONT1 ($2P_{\text{ONT1}}$) and ONT2 ($2P_{\text{ONT2}}$), respectively. Since ones and zeros are equally distributed in the signal, the corresponding average optical power levels are P_{ONT1} and P_{ONT2}, also indicated as horizontal dashed lines in Fig. 3.24(b). If the DCA is triggered on the data frequency of 1.25 GHz, Fig. 3.24(d-f), then the eye diagram of the data signal from both ONT1 and ONT 2 becomes visible.

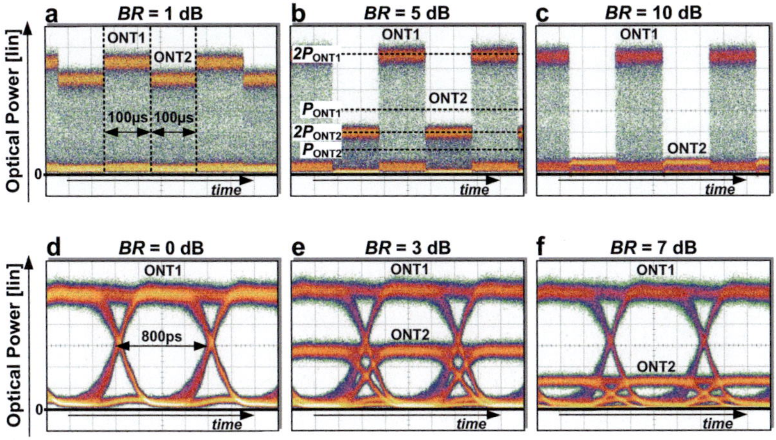

Fig. 3.24: Output signal of the rival TDM-GPON burst mode transmitter of Fig. 3.23(c) for different burst ratios *BR*. (a-c): If triggered on the burst frequency of 5 kHz, the 100 µs time slots of ONT1 and ONT2 become visible. (b) Horizontal dashed lines indicate the average optical powers P_{ONTn} and the high power levels $2P_{\text{ONTn}}$ for transmitted ones. (d-f): If triggered on the 1.25 Gbit/s NRZ-OOK data signal, the eye diagrams of ONT1 and ONT2 can be seen. (d) BR = 0 dB corresponds to the output signal of the rival WDM-channel transmitter of Fig. 3.23(c).

RoF Receiver

Since we want to investigate the impact of the rival signals on the RoF1 signal, a 2 nm bandpass filter centered around $\lambda_{\text{RoF1}} = 1300$ nm is used to select only the RoF1 signal λ_{RoF1} in the RoF Rx, Fig. 3.23(e). This filter also reduces the ASE noise from the SOA under test. An Alphion SOA (model SAO29p, 26 dB FtF gain, 3 dB saturation input power of −13 dBm) is employed as Rx preamplifier. A subsequent 2 nm bandpass filter (not shown in Fig. 3.23(e)) reduces the ASE noise added by the Rx preamplifier. After photodetection with a PIN photodiode, the RF signal is amplified by a low noise amplifier (LNA) and input to a vector signal analyzer (VSA) that demodulates the received RoF1 signal. We measure the error vector magnitude (EVM) of the RoF1 signal to evaluate the impact of the rival signals on the RoF1 signal. Conditions for error-free (BER = 10^{-9}, i.e. EVM = 16.7 % for QPSK [158]) signal amplification are then determined as a function of the three rival signal scenarios.

The optical input powers into the preamplifier Rx and on the photodiode are always kept constant in order to have the same Rx conditions for all SOA operating points. Several aspects have to be taken into account when choosing the optimum optical Rx input power P_{RX} which should provide a bit error ratio (BER) < 10^{-9} of the RoF1 signal in a back-to-back measurement without the SOA under test:

First, the Rx preamplifier SOA should be operated in its linear regime to avoid nonlinear effects at the Rx. Second, P_{RX} should be sufficiently small to allow the operation of the SOA under test at very low input powers, which enables the investigation in a large optical power range. However, smaller P_{RX} require a larger minimum modulation depth m_{RoF1} to achieve a BER = 10^{-9}, which in turn would decrease the RF power range that can be investigated.

A B2B measurement without the SOA under test is conducted to find a P_{RX} that meets these requirements. The grey shaded area in the inset of Fig. 3.23(e) shows the pairs of $(m_{\text{RoF1}}, P_{\text{RX}})$ where a BER < 10^{-9} is achieved in the back-to-back case. We choose $P_{\text{RX}} = -17.5$ dBm which is 4.5 dB below the 3 dB input saturation power of the Rx SOA. Furthermore, since the SOA under test has an unsaturated FtF gain of 18 dB, we can investigate SOA input powers down to −35.5 dBm before we enter the region BER > 10^{-9} of the RoF1 Rx. However, for $P_{\text{RX}} = -17.5$ dBm, an optical modulation depth m_{RoF1} > 2% is needed for a BER < 10^{-9}.

3.3.2 Amplification of a RoF Signal without Rival Signals

The amplification of a RoF signals without rival signals was investigated in *Section 3.2.3* for the QD SOA, and the OPDR and EPDR were determined. In this *Subsection*, we perform the same experiment for the Alphion QW SOA (see *Section 3.2.1*), which will be used in the experiments with the rival signals in the next *Subsections*. There, the measurement results from this *Subsection* without the impact of a rival signal will serve as a reference.

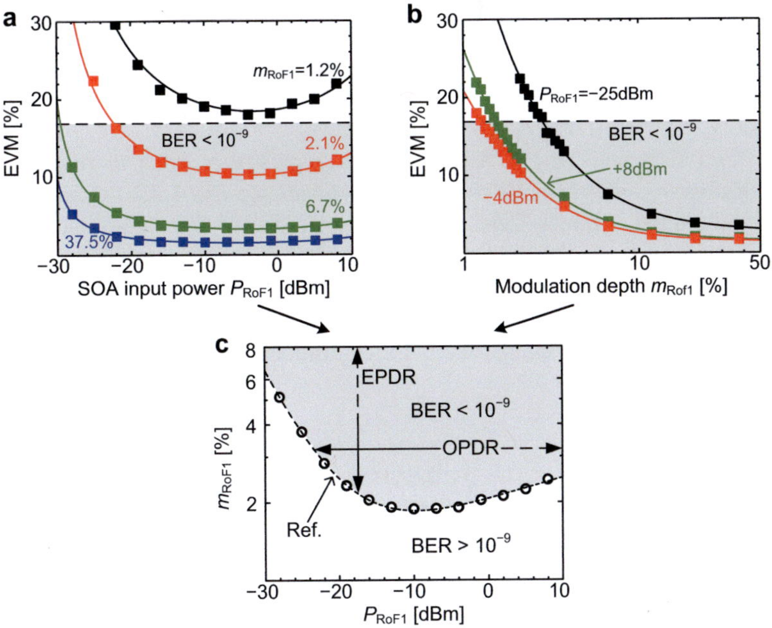

Fig. 3.25: (a) Error vector magnitude (EVM) of the RoF1 signal without any rival signal as a function of the average SOA input power P_{RoF1} for different optical modulation depths m_{RoF1}. (b) EVM as a function of m_{RoF1} for different P_{RoF1}. The optimum operation point is the 3 dB input saturation power (−4 dBm). (c) (P_{RoF1}, m_{RoF1})-pairs for error-free signal amplification (BER < 10^{-9}, grey shaded area) as obtained from (a,b). The border line (Ref.) of the region BER < 10^{-9} serves as a reference in measurements with rival signals. The available OPDR for a given m_{RoF1} and the available EPDR for a fixed P_{RoF1} with which error-free signal amplification for the RoF1 signal is obtainable are indicated. The EPDR can be optimized at the expense of OPDR and vice versa.

Fig. 3.25(a) shows the EVM of the RoF1 signal (20 MBd QPSK) as a function of the average SOA input power P_{RoF1} for different modulation depths m_{RoF1}. The horizontal dashed line indicates the EVM limit for QPSK corresponding to a BER = 10^{-9}. The grey shaded area marks the region where BER < 10^{-9}. As discussed in *Section 3.2.3*, for low P_{RoF1}, the signal quality is limited by noise and for large P_{RoF1}, the SOA is operated in saturation leading to higher EVM. For m_{RoF1} = 6.7%, a large OPDR > 40 dB is found. The optimum operating point (minimum EVM value irrespective of m_{RoF1}) is found to be around the 3 dB input saturation power of −4dBm. Fig. 3.25(b) shows the EVM as a function of m_{RoF1} for different P_{RoF1}. For P_{RoF1} = −4 dBm, a large EPDR >30 dB is found. In Section 3.2.3, the EPDR was defined as the *RF power range* within which error-free signal amplification is obtained for a given P_{RoF1}. The EPDR expressed in dB can be calculated from the optical modulation depth m_{RoF1} by relating upper and lower limit of m_{RoF1} for error-free signal amplification by $20\log_{10}(m_{\mathrm{RoF1,upper}} / m_{\mathrm{RoF1,lower}})$.

From the results of Fig. 3.25(a,b) ($P_{\mathrm{RoF1}}, m_{\mathrm{RoF1}}$)-pairs for error-free signal amplification without rival signal are obtained, see Fig. 3.25(c). The error-free operation region is indicated by the grey shaded area. The border line (Ref.) of the region BER < 10^{-9} in Fig. 3.25(c) will serve as a reference in measurements with rival signals. Fig. 3.25(c) indicates furthermore the available OPDR for a given m_{RoF1} and the available EPDR for a fixed P_{RoF1} with which error-free signal amplification for the RoF1 signal is obtainable.

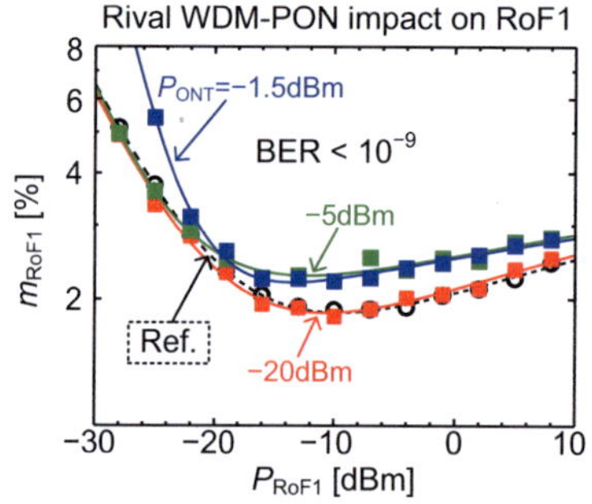

Fig. 3.26: Influence of the continuous NRZ-OOK WDM-PON on the RoF1 signal. Only little influence is observed compared to the RoF1 reference (Ref.). Even in deep saturation of the SOA ($P_{\mathrm{ONT}} = -1.5$ dBm) where cross-gain modulation takes place, the RoF1 is hardly affected by the WDM-PON signal, since both RF spectra do not coincide (NRZ-OOK < 1.25 GHz, $f_{\mathrm{RoF1}} = 5$ GHz). Modified from [C23].

3.3.3 Amplification of a RoF Signal and a Rival WDM-PON Signal

In Fig. 3.26, the effect of a WDM-PON operated with a continuous 1.25 Gbit/s NRZ-OOK signal (ONT) onto the RoF1 test channel is studied. For weak ONT power levels ($P_{\mathrm{ONT}} = -20$ dBm), no difference to the reference scenario (Ref.) without WDM-PON is observed. Even for a large ONT power level ($P_{\mathrm{ONT}} = -1.5$ dBm) where the SOA is deeply saturated (cross-gain modulation (XGM) could be expected), only little influence on the test channel is observed, since in our case the RF spectra of both signals do not coincide (NRZ-OOK < 1.25 GHz, $f_{\mathrm{RoF1}} = 5$ GHz and a bandwidth of 20 MHz). Consequently, the RoF1 channel is hardly affected by the presence of a WDM-PON subscriber.

3.3.4 Amplification of a RoF Signal and a Bursty TDM-GPON

Next, the influence of the bursty TDM-GPON-like signal is studied against the average optical input power P_{ONT1} into the SOA and the burst ratio *BR*. Different P_{ONT1} represent different minimum distances of the nearest ONT1 of the TDM-GPON to the SOA and/or different split ratios of the passive splitters, whereas different *BR* correspond to displacements of ONT2 with respect to ONT1.

The impact of a bursty TDM-GPON-like signal is investigated for the extreme case of a weak RoF1 signal ($P_{\mathrm{RoF1}} = -20$ dBm). Fig. 3.27(a,b,c) show the bursty TDM GPON-like signal ($\lambda_{\mathrm{TDM}} = 1310$ nm) *after* amplification by the SOA for different P_{ONT1} (−20 dBm, −5 dBm, and 1.5 dBm) and a fixed $BR_{\mathrm{Tx}} = 5$ dB between ONT1 and ONT2 (before amplification). The DCA is triggered on the burst frequency of 5 kHz.

Since P_{RoF1} is weak, the SOA operating point is determined by the TDM-GPON-like signal. In Fig. 3.27(d,e,f), the operating point for the different P_{ONT1} is indicated on the 1310 nm gain curve of the SOA under test by horizontal dashed lines which mark the high power levels $2P_{\mathrm{ONT1}}$ of ONT1 and $2P_{\mathrm{ONT2}}$ of ONT2, respectively. The ratio $2P_{\mathrm{ONT1}}/2P_{\mathrm{ONT2}}$ corresponds the $BR_{\mathrm{Tx}} = 5$ dB. For $P_{\mathrm{ONT1}} = -20$ dBm, the SOA is operated in the linear region, Fig. 3.27(d), and both the signals from ONT1 and ONT2 are amplified with the unsaturated SOA gain G_0. For $P_{\mathrm{ONT1}} = -5$ dBm, the SOA is operated on the edge of saturation, Fig. 3.27(e), and for $P_{\mathrm{ONT1}} = -1.5$ dBm the SOA is operated in deep saturation, Fig. 3.27(f). As a consequence, the burst level of ONT1 experiences less gain (due to the SOA gain compression) than the burst level of ONT2. Thus the burst ratio after the SOA BR_{Rx} decreases, see Fig. 3.27(a,b,c).

The impact on the RoF1 carrier ($\lambda_{\mathrm{RoF1}} = 1300$ nm) is shown in Fig. 3.27(g,h,i), where the received RoF1 signal (triggered on the burst frequency) is depicted. If the SOA is operated in the linear region, Fig. 3.27(g), no effect of the bursty TDM-GPON-like signal on the RoF1 signal is observed. For $P_{\mathrm{ONT1}} = -5$ dBm, the burst level of ONT1 drives the SOA into gain saturation and this way the 1.25 Gbit/s signal is transferred by means of XGM onto λ_{RoF1}, cf. Fig. 3.27(h). On the other hand, an ONT2 burst operates the SOA still near the linear region and thus no XGM is visible. For

$P_{\mathrm{ONT1}} = -1.5$ dBm, the burst level of both ONTs drive the SOA into saturation, thus XGM is observed for both ONTs, Fig. 3.27(i). Note that XGM is an inverting effect: In the presence of a strong ONT signal, λ_{RoF1} experiences less gain.

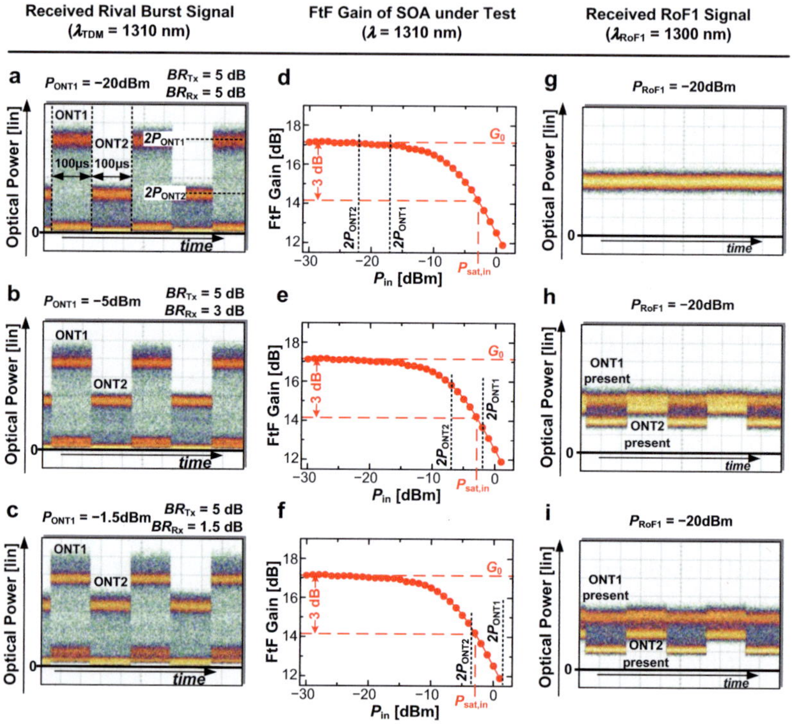

Fig. 3.27: Impact of a bursty TDM-GPON-like signal (1.25 Gbit/s NRZ-OOK, burst length 100 µs) on a weak RoF1 signal ($P_{\mathrm{RoF}} = -20$ dBm) when simultaneously amplified by the SOA. (a,b,c) Rival burst signal at $\lambda_{\mathrm{TDM}} = 1310$ nm after amplification by the SOA for different average ONT1 power levels P_{ONT1} (recorded with a DCA and triggered on the burst frequency of 5 kHz). The burst ratio BR_{Tx} between ONT1 and ONT2 before amplification is always 5 dB. (d,e,f) Different operating points on the SOA gain curve for the different ONT power levels P_{ONT1} and P_{ONT2}, respectively. (g,h,i) Impact on the RoF1 signal at $\lambda_{\mathrm{RoF1}} = 1300$ nm after amplification by the SOA (also triggered on the burst frequency). For linear operation, the burst ratio after amplification $BR_{\mathrm{Rx}} = BR_{\mathrm{Tx}}$(a), and no impact of the bursty TDM signal on the RoF1 signal is observed (g). If the burst level of one or both ONTs reaches the nonlinear SOA regime, then $BR_{\mathrm{Rx}} < BR_{\mathrm{Tx}}$ due to SOA gain compression (b,c). Cross-gain modulation (XGM) copies the (inverted) burst signal onto λ_{RoF1}. However, the major impact on the RoF1 signal occurs at the transition between the two bursts as can be seen in Fig. 3.28(a). SOA overshoots of the 1.25 Gbit/s data signal are visible in deep saturation (c, ONT1).

Fig. 3.28(a) depicts the RF signal power of the received 20 MBd QPSK signal of RoF1. The major impact on the RF signal is seen during the transition from the signal of ONT1 to the signal of ONT2 (see middle constellation diagram). While there is no transition, the situation corresponds to the case of a rival continuous WDM-PON signal where the RoF1 signal is hardly affected, as discussed in *Section* 3.3.3. Fig. 3.28(b) shows the RF power fluctuation ΔP_{RF1} of the received RoF1 signal as a function of the input burst ratio for different P_{ONT1}. The RF power fluctuation ΔP_{RF1} increases under SOA gain suppression for an increasing burst ratio. No power fluctuation occurs, if the SOA is operated in the linear region (see $P_{ONT1} = -20$ dBm in Fig. 3.28(b)), which is recommended to avoid a permanent impact of the bursty signal on the RoF1 signal.

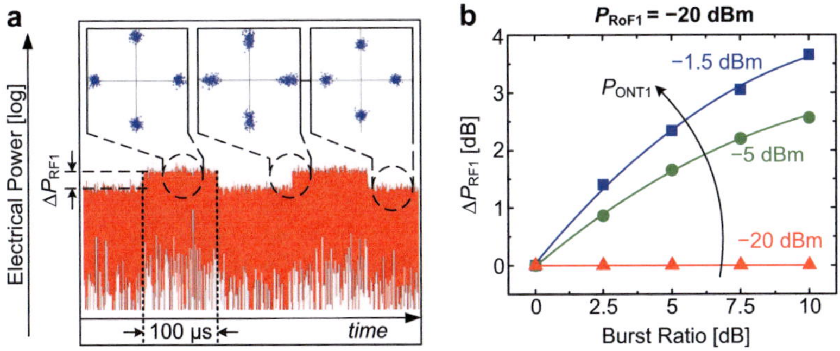

Fig. 3.28: Impact of the bursty TDM-GPON-like signal seen as RF power fluctuation ΔP_{RF1} on the received RoF1 signal in the baseband. The insets show constellation diagrams (QPSK) during an ONT1 burst (left), during an ONT2 burst (right), and during a transition from ONT1 to ONT2 bursts (middle). (d) The power fluctuation ΔP_{RF1} depends on the burst ratio *BR* and the optical power P_{ONT1} of ONT1. Modified from [C23] © 2011 The Optical Society.

3.3.5 Amplification of a RoF Signal and a Rival RoF Signal

We study the impact of the rival RoF2 onto RoF1 when varying the average optical power P_{RoF1} and P_{RoF2}. This corresponds to a scenario where RoF WAPs of subscribers 1 and 2 are located at different distances to the SOA, see Fig. 3.22. In addition we vary the optical modulation depths m_{RoF1} and m_{RoF2} to emulate different RF power levels at the WAP originating from different mobile users who have different distances to the WAP.

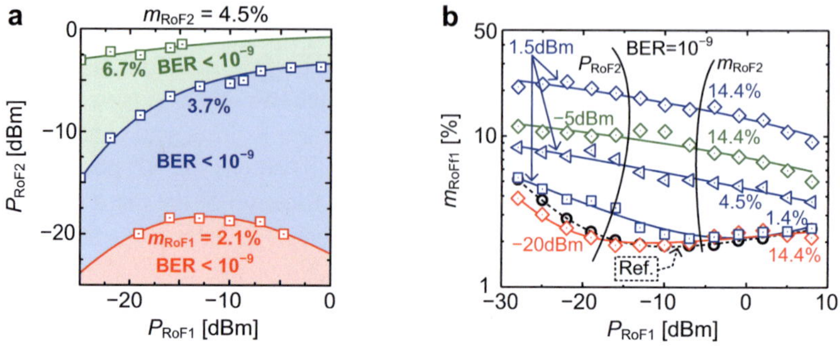

Fig. 3.29: Impact of a rival RoF2 signal (λ_{RoF2}= 1310 nm, f_{RoF2} = 5 GHz) on the RoF1 signal (λ_{RoF1}= 1300 nm, f_{RoF1} = 5 GHz). (a) (P_{RoF1}, P_{RoF2})-pairs with which error-free amplification (indicated by the area below the fitted curves which correspond to BER = 10^{-9}) is guaranteed. (b) (P_{RoF1}, m_{RoF1})-pairs for error-free signal amplification (indicated by the area above the fitted curves which correspond to BER = 10^{-9}) of the RoF1 signal as a function of P_{RoF2} and m_{RoF2} of the rival RoF2 signal. For P_{RoF2} = −5 dBm and P_{RoF2} = −1.5 dBm, the SOA is operated in saturation irrespective of P_{RoF1} → transfer of the rival signal RoF2 from λ_{RoF2} onto λ_{RoF1} by means of XGM. This leads to degradations of the RoF1 signal, since both wireless signals have same carrier frequency ($f_{RoF1} = f_{RoF2}$) and thus superimpose in the RF spectrum. The strength of degradation is determined by m_{RoF2}. Modified from [C23] © 2011 The Optical Society.

Fig. 3.29(a) shows for given RoF modulation depths m_{RoF1} and m_{RoF2} pairs of acceptable power levels (P_{RoF1}, P_{RoF2}) with which error-free amplification is guaranteed. The error-free region is *below* the curves and increases with increasing m_{RoF1}. In practice, distortions due to rival RoF2 can be overcome by reducing the power of RoF2 channel before feeding it into the SOA, i.e., by adding some additional attenuation at the WDM multiplexer.

Fig. 3.29(b) shows more results from Fig. 3.29(a) but plotted for pairs of (P_{RoF1}, m_{RoF1}) for various P_{RoF2} and average modulation depths m_{RoF2}. The regions *above* the fitted curves show error-free signal amplification BER < 10^{-9}. The dashed reference (Ref.) is the case without rival signal. The plots show that the larger is P_{RoF2} and the larger is m_{RoF2} the larger must be m_{RoF1} for an error-free signal amplification. For a stronger rival signal (P_{RoF2} = −5 dBm, −1.5 dBm), the SOA is operated in saturation for any P_{RoF1}. In this operating regime the rival RoF2 signal is transferred from λ_{RoF2} = 1310 nm onto λ_{RoF1} = 1300 nm by means of XGM. Due to the identical RF carrier frequencies $f_{RoF1} = f_{RoF2} = 5$ GHz both wireless signals superimpose in the RF spectrum which leads to severe signal degradations. The strength of degradation is determined by the modulation depth m_{RoF2}. The impact of RoF1 on RoF2 for stronger m_{RoF1} is not discussed in this study.

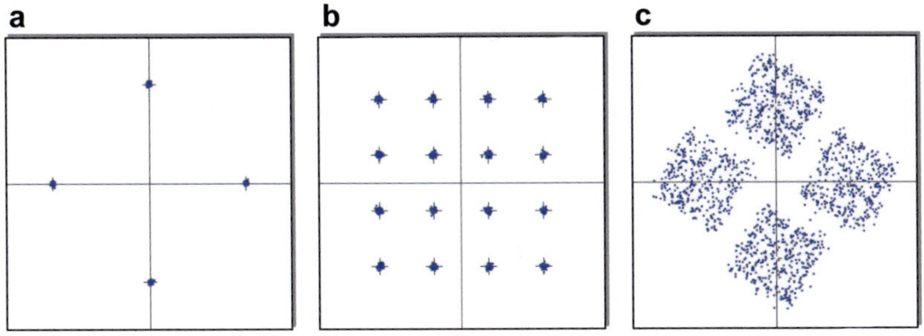

Fig. 3.30: Typical QPSK (a) and 16QAM (b) constellation of the RoF1 signal and the rival RoF2 signal when the SOA is operated properly and no signal distortions occur. (c) Strong RoF1 signal degradations due to cross-gain modulation (XGM) when the SOA is operated in deep saturation. The rival 16QAM RoF2 signal seems to be mixed with the QPSK RoF1 signal.

However, it is obvious that the modulation depths and optical power levels of the different radio transmitters have to be regulated to achieve an acceptable signal quality of both signals after the SOA. The XGM efficiency and therewith the signal degradation of the RoF1 signal are mitigated if P_{RoF1} is increased. As a consequence we derive that impact of RoF1 against RoF2 can be minimized by keeping the sum of optical powers of all RoF subscribers below the SOA saturation power level.

Fig. 3.30(a,b) show typical constellation diagrams of the received RoF1 signal with QPSK (a) and the rival RoF2 signal with 16QAM (b), if the SOA is operated properly and no signal distortions occur. Fig. 3.30(c) shows the constellation diagram of the received RoF1 signal, when the SOA is operated in deep saturation and severe distortions due to XGM lead to the degradation of the RoF1 signal. The constellation diagram indicates that the QPSK modulation of the RoF1 signal is mixed with the 16QAM modulation of the rival RoF2 signal.

It should be noted again that we discussed the case where both RF signals use the same carrier frequency. Thus their RF spectra are superimposed after the XGM process. Choosing slightly different carrier frequencies for both RF signals may solve this problem. In practice, however, when a large number of RoF subscribers are present, the available RF spectrum rapidly is used up and thus limiting the number of RoF subscribers. A carrier frequency management system becomes necessary monitoring which RF carrier is assigned to which RoF subscriber and is amplified by which SOA in the network.

3.3.6 Conclusions

In the linear SOA operating region, no impact of the rival signals on the RoF signal is observed. Major impacts of a rival RoF and a bursty TDM-GPON-like signal on a RoF signal in a SOA reach extended converged RoF and WDM-TDM-GPON were shown to occur in gain saturation as a consequence of XGM. Thus, the nonlinear operating region of the SOA should be avoided and the sum of all powers from all subscribers at the SOA input should be kept below the SOA saturation input power to guarantee a good signal quality after amplification.

3.4 SOA with RoF OFDM Signals

As mentioned in *Section 1.1.2*, orthogonal frequency division multiple access (OFDMA) is currently discussed for PONs in optical access networks [12, 13]. Furthermore, OFDMA is widely used in wireless broadband applications ranging from wireless local area networks to 4G LTE-based mobile communications systems [159]. In view of the fact, that SOAs are a key device in the context of both optical-wireless and metro-access network convergence, an investigation of SOA with OFDM signals is necessary.

In *Subsection 3.4.1*, we investigate the amplification of a broadband RoF OFDM signal by a single SOA (we use the QD SOA from *Section 3.2.1*). In particular, we study the influence of the modulation response of the SOA on a real-valued electrical OFDM signal, which is encoded with intensity modulation on an optical carrier.

In *Subsection 3.4.2*, the same RoF OFDM signal is transmitted over a cascade of SOAs to mimic the application scenario of a converged metro-access ring network as illustrated in Fig. 1.7. We perform experiments in a recirculating loop (17 km) with the QD SOA from *Section 3.2.1* and we investigate total propagation distances of up to 135 km. We give rules for the range of optical channel power of a cascade of SOAs to guarantee the functionality of a converged metro-access network comprising several ROADMs.

3.4.1 Single SOA with OFDM signals

Experimental Setup

The experimental setup shown in Fig. 3.31 is used to study the amplification of a RoF OFDM signal by a single SOA (QD SOA from *Section 3.2.1*). In the transmitter (Tx), a continuous wave (CW) laser ($\lambda = 1295$ nm) is externally modulated by a Mach-Zehnder modulator (MZM) biased at its quadrature point (intensity modulation, IM).

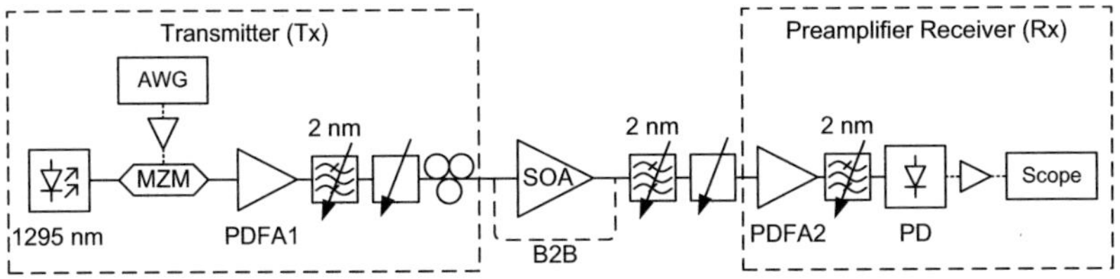

Fig. 3.31: Experimental setup to investigate the amplification of RoF OFDM signal by a single SOA. AWG=arbitrary waveform generator, PDFA=praseodymium-doped fiber amplifier, B2B=back-to-back, PD=photodiode.

A 5 GHz-wide real-valued *electrical* OFDM signal is offline generated with MATLAB and physically generated by a Tektronix 7102B arbitrary waveform generator (AWG) at a sampling rate of 20 GSa/s. The size of the inverse fast Fourier transform (IFFT) is 256 leading to a subcarrier spacing of 78.125 MHz. The OFDM signal comprises 63 subcarriers: One unmodulated subcarrier located at the intermediate frequency of 2.5 GHz (center frequency), 4 BPSK-encoded pilot tones equidistantly distributed over the OFDM spectrum, and 58 subcarriers modulated with data having all identical modulation format. We investigate an OFDM-BPSK signal and an OFDM-QPSK signal with a line rate of 4.53 Gbit/s and 9.06 Gbit/s, respectively. A cyclic prefix (CP) leads to an overhead of 25 % and results in a reduced data rate of 3.63 Gbit/s and 7.25 Gbit/s, respectively.

The electrical peak-to-average power ratios (PAPR) of the OFDM-BPSK and OFDM-QPSK signals at the AWG's output are 19.8 and 24.8, respectively. An electrical driver amplifier is used to boost the OFDM signal before the MZM. After the driver amplifier at the MZM RF port, the root mean square (rms) value of the OFDM signal is $V_{\text{rms}} \approx 0.4$ V and the PAPR of the OFDM-BPSK and OFDM-QPSK signal is 12.1 and 13.4, respectively. We define the optical modulation depth as $m = \sqrt{2}\,V_{\text{rms}} / V_{\pi}$. With $V_{\pi} = 5.5$ V for the MZM, a modulation depth of $m_1 = 10$ % around the quadrature point results after the Tx.

To investigate also large SOA input powers, the RoF OFDM signal is boosted by a praseodymium-doped fiber amplifier (PDFA1). Since the QD SOA gain is strongly polarization dependent, the QD SOA input polarization is adjusted with a polarization controller. The input power to our preamplifier receiver (Rx) is kept constant at −15 dBm which is well above its sensitivity threshold for a bit error ratio (BER) of 10^{-9}. A second PDFA2 amplifies the received signal and a pin-photodetector (PD) with an electrical bandwidth of 20 GHz directly detects (DD) the optical signal, which is kept at a constant average power level of +5 dBm. We use optical bandpass filters (bandwidth 2 nm) to remove out-of-band noise after both PDFAs and the SOA under test. The back-to-back (B2B) case without amplification by the SOA is also indicated in Fig. 3.31.

The received OFDM signal is electrically amplified and then recorded with a Tektronix DPO70804B real-time digital oscilloscope for subsequent offline processing and signal analysis. We use the custom OFDM demodulation tool of the commercial Agilent VSA software to demodulate and analyze our OFDM signal.

Results

In this *Subsection*, we qualitatively evaluate the results of the amplification of the broadband OFDM-QPSK signal by a single SOA to identify operation limits at different SOA input power levels. A more quantitative evaluation in terms of the error vector magnitude (EVM) and optical power dynamic range (OPDR) is given in the next *Subsection*.

Table 3.4 shows the constellation diagrams of the different subcarriers of the OFDM-QPSK signal after amplification by the SOA for the SOA input powers −25 dBm, −10 dBm, and +10 dBm. Furthermore, the B2B constellations without amplification by the SOA are shown. Each presented constellation diagram includes the demodulated QPSK symbols of seven OFDM subcarriers. The QPSK symbols of different OFDM subcarriers are indicated by different colors (from blue to gray). The first column indicates which OFDM subcarriers are overlaid in the constellation diagram in the respective row and which frequency range is covered by the different OFDM subcarriers. Note that the pilot tones (subcarriers 8, 24, 40, and 56) and the unmodulated center carrier (subcarrier 32) are not shown. For measured constellation diagrams in Table 3.4 the software equalizer function which corrects for linear impairments in the signal path (channel frequency response) during offline processing at the Rx was turned off. This makes the influence of the SOA modulation response directly visible in the constellation diagram.

In the B2B measurement, phase rotations in the frequency range between 80 MHz and 550 MHz can be observed. The phase rotation is caused by the RF amplifiers in front of the MZM and after the PD. However, the demodulated QPSK symbols of the subcarriers 1…7 (different colors) are clearly distinguishable. For larger subcarrier numbers (higher frequencies), the QPSK constellation diagrams show a very good signal quality and no phase rotations are observed. Yet, the low-pass frequency response of the AWG and the RF amplifiers becomes visible when moving towards larger frequencies where the magnitude of the received symbols is smaller than at lower frequencies.

The constellation diagrams for an SOA input power of −25 dBm (linear SOA operating region) show a symmetric broadening of the constellation points. The performance in this SOA operating region is limited by amplified spontaneous emission (ASE) noise. The phase rotation at low frequency is preserved from the B2B case.

For an SOA input power of −10 dBm (3 dB below the 1 dB input saturation power), the signal quality is comparable to the B2B measurement. However, the SOA modulation response, see *Section 3.2.1*, starts to affect the OFDM signal. The high-pass characteristic of the SOA modulation response counteracts the low-pass characteristic which was observed in the B2B measurement. This leads to a uniform magnitude over all OFDM subcarriers.

For a large SOA input power of +10 dBm, the SOA is operated in deep saturation (nonlinear operating region). Here, the subcarriers 1…15 are strongly attenuated due to the high-pass characteristic of the SOA modulation response, see *Section 3.2.1*. Besides, a similar symmetric broadening of the constellation points is observed as for an input power of −25 dBm where ASE noise limits the performance. However, for large SOA input powers nonlinear interactions of the OFDM subcarriers (inter-subcarrier mixing) limit the performance. As a consequence of the nonlinear effects and the high-pass characteristic of the SOA modulation response, the subcarriers are stronger degraded at smaller frequencies than at larger frequencies.

OFDM subcarrier number (frequency range)	Optical B2B	SOA input power = −25 dBm	SOA input power = −10 dBm	SOA input power = +10 dBm
1…7 (0.08…0.55 GHz)	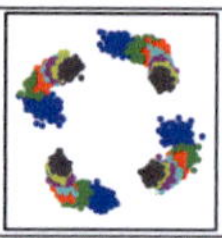			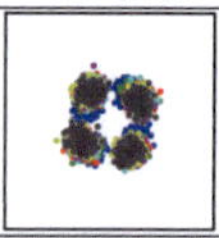
9…15 (0.70…1.17 GHz)	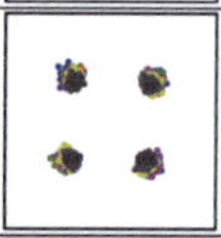	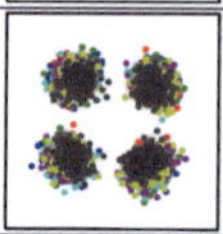	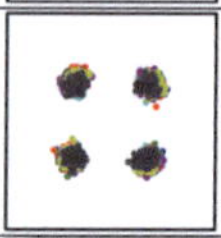	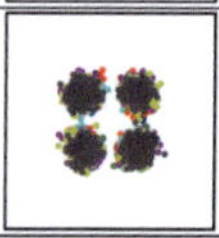
16…22 (1.25…1.72 GHz)		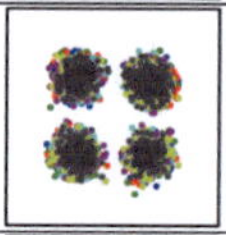		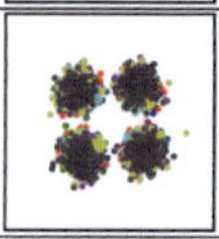
23, 25…30 (1.80…2.34 GHz)	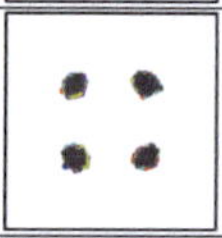	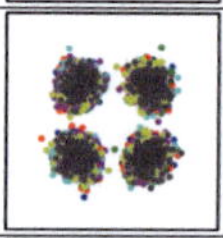	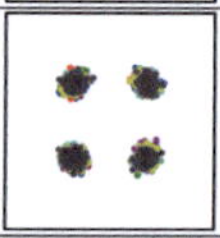	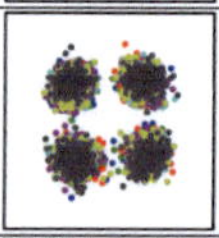
31, 33…38 (2.42…2.97 GHz)		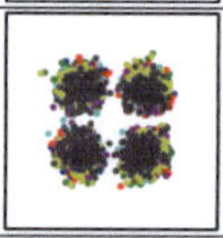		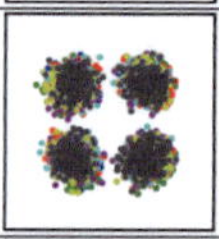
39, 41…46 (3.05…3.59 GHz)	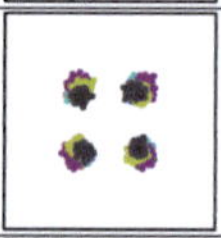	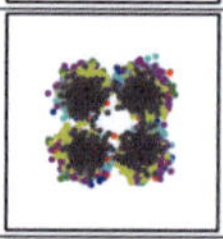		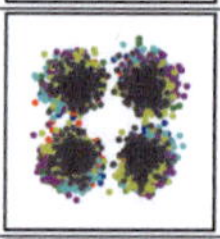
47…53 (3.67…4.14 GHz)	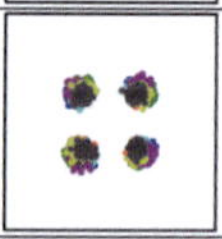	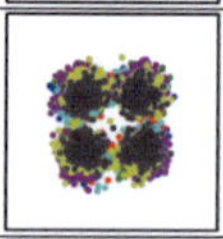		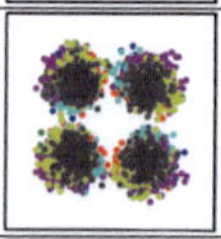
54, 55, 57, 59…61, 63 (4.22…4.92 GHz)		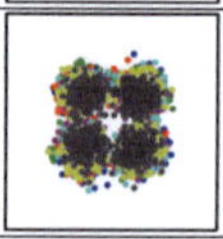	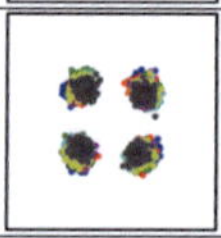	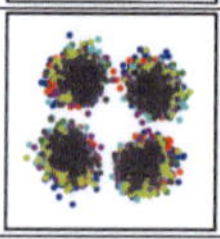

Table 3.4: Constellation diagrams of the OFDM-QPSK subcarriers after amplification by a single SOA for the SOA input powers −25 dBm (linear operation), −10 dBm (3 dB below the 1 dB input saturation power), and +10 dBm (deep saturation). The first column indicates which OFDM subcarriers are overlaid in the constellation diagrams and which frequency range is covered. The second column shows the back-to-back (B2B) constellations without SOA. Strong signal degradations in saturation result from inter-subcarrier mixing and the high-pass characteristic of the SOA modulation response. The pilot tones (subcarriers 8, 24, 40, and 56) and the unmodulated center carrier (subcarrier 32) are not shown.

3.4.2 SOA Cascade with OFDM signals

Experimental Setup

To investigate the amplification of the broadband RoF OFDM signals in a converged metro-access ring network scenario as depicted in Fig. 1.7, we modify our experimental setup of Fig. 3.31. The setup in Fig. 3.32(a) includes a recirculating loop (RL) (a detailed experimental setup chart is given in *Appendix A.6*). The aim is to investigate a subchannel performance as a function of the input power of the resulting SOA cascade and the possible span for the broadband RoF OFDM data signals.

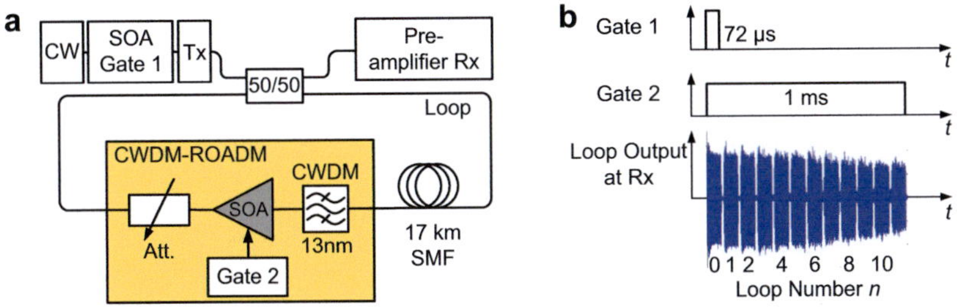

Fig. 3.32: Recirculating loop setup. (a) The loop comprises an SOA, an attenuator, a CWDM filter and 17 km of SMF fiber. OFDM-BPSK/QPSK signals are generated which are adjusted in power before launching them into the first SOA in the recirculating sequence. The receiver detects the signal passing *n*-times through the SOA. (b) Transmitter and QD SOA gate window and loop output signal at the Rx.

The following modifications with respect to the experimental setup of Fig. 3.31 have been made: The Tx laser is gated with an SOA, Gate 1 time = 72 µs, Fig. 3.32(b). In this time period, the OFDM-BPSK or the ODFM-QPSK introduced in *Section 3.4.1* is encoded on the optical carrier. The 72 µs OFDM burst is then launched into the recirculating loop via a 3 dB coupler. The loop comprises the 3 dB coupler, 17 km of standard single mode fiber (SMF, loop propagation time 85 µs), and the ROADM which is equipped with a CWDM filter (13 nm bandwidth), a gated (Gate 2) SOA (same QD SOA as in *Section 3.4.1*), and a loop attenuator (Att.) to mimic ROADM losses. The Gate 2 time = 1 ms and Gate 2 opens simultaneously with Gate 1, see Fig. 3.32(b). The QD SOA gain polarization dependence (10 dB) is accounted for by a polarizer and polarization controllers. The loop losses are adjusted once by the Att. to compensate the QD SOA small-signal gain. The OFDM output signals of the loop cycles are received with the pre-amplifier Rx of Fig. 3.31. The OFDM signal quality is evaluated in terms of error vector magnitude (EVM) (*Appendix A.3.2*, [J6]). Note that the first recorded OFDM burst (loop number 0 in Fig. 3.32(b)) corresponds to the OFDM burst which directly comes from the Tx and which does not enter the loop. Since the Gate 2 of the SOA under test opens at the same time as Gate 1 at the Tx, ASE noise from the SOA under test is loaded from the very beginning to the loop and then amplified by the SOA together with the first OFDM burst entering the loop (loop number 1 in Fig. 3.32(b)).

Results

In Fig. 3.33 the achievable number of loop cycles for a range of channel input power entering the first SOA is presented for the OFDM-BPSK and OFDM-QPSK signals. Fig. 3.33(a) indicates that for OFDM-BPSK 8 cycles (136 km) are achievable with an EVM corresponding to a BER better than 10^{-3} (horizontal dashed line). The EVM value is calculated from all OFDM subcarriers. In contrast, Fig. 3.33(b) indicates that only 4 loops (68 km) are achievable with OFDM-QPSK modulation.

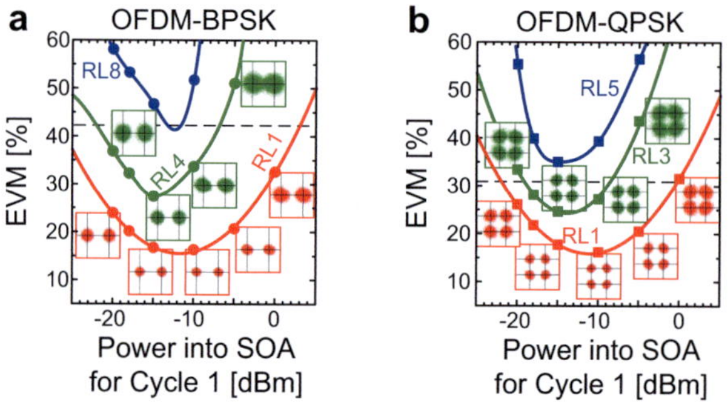

Fig. 3.33: EVM measurement for “cascaded” SOAs in a recirculating loop (RL) transporting OFDM signals. EVM averaged over all subcarriers vs. input power entering the SOA at the first loop cycle for different loop cycle numbers and modulation formats. (a) 4.5 Gbit/s OFDM-BPSK, (b) 9 Gbit/s OFDM-QPSK with some constellations as insets.

In Fig. 3.34 the maximum and minimum input power levels (optical power dynamic range given that BER $\leq 10^{-3}$) at the SOA input at the first loop cycle are presented as a function of the number of recirculating loop cycles, i.e., the number of cascaded SOAs. The upper input power limit is given by nonlinear interaction of the OFDM subcarriers inside the SOA (intermodulation) and by the high-pass characteristic of the SOA (suppression of modulation depth at low-frequency subcarriers), as explained in *Section 3.2.2*. The lower input power limit is set by the OSNR and monotonously increases for increasing loop numbers for maintaining a sufficient OSNR. The upper power limit monotonously decreases for increasing loop cycles. The optimum SOA input power lies around 5 dB below the 1 dB saturation input power (•).

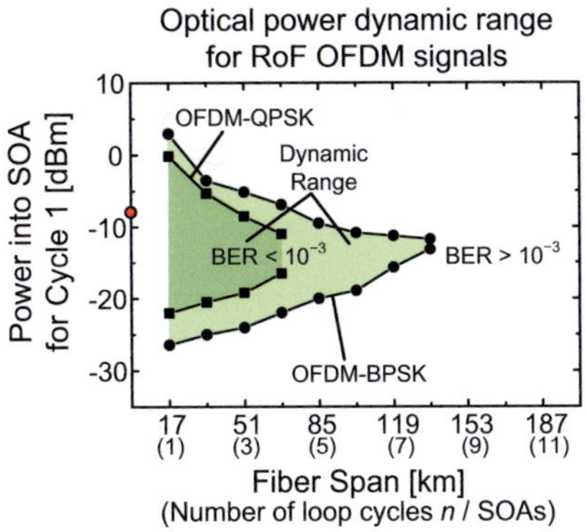

Fig. 3.34: Optical power dynamic range for RoF OFDM-BPSK/QPSK signals for a corresponding BER = 10^{-3} as a function of the fiber span (number of loop cycles / cascaded SOAs). An increase in modulation format complexity from OFDM-BPSK to OFDM-QPSK reduces the reach from around 136 km to around 68 km. The optimum input power to the first SOA lies about 5 dB below the 1 dB saturation input power (•).

3.4.3 Conclusions

Based on our experimental results, the following rules (1…5) apply for (cascaded) SOAs in a converged metro-access network with RoF OFDM signals:

1) The OFDM signal limitations can be attributed to gain saturation which induces subcarrier mixing and the attenuation of low-frequency subcarriers.
2) Lower-order modulation formats occupying the same bandwidth are preferable to achieve large span numbers. BPSK tolerates a higher EVM compared to QPSK (identical BER), and thus it has a lower OSNR requirement.
3) The dynamic range reduces significantly for an increasing number of cascaded SOAs, which limits the acceptable power variations that are introduced, e.g., in GPON upstream paths.
4) The best input power to the first SOA lies about 5 dB below the 1 dB saturation input power, showing that linearity requirements are high.
5) OFDM signals seem to be well suited for short-range connections.

4 High-Capacity mm-Wave Wireless Links in Optical Access Networks

In this *Chapter* we investigate the insertion of a multi-gigabit mm-wave wireless link into a fiber-optic environment. Different application scenarios where high-speed mm-wave wireless links may be employed in optical access networks have been introduced in *Section 1.4*.

In *Section 4.1*, the feasibility of a wireless link at 220 GHz based on electronic up- and down-conversion is demonstrated to connect two optical links at data rates of up 20 Gbit/s. Either non-return-to-zero (NRZ) on-off keying (OOK) with data rates up to 20 Gbit/s, or electrical orthogonal frequency division multiplexing (OFDM) with data rates up to 9 Gbit/s is used. The wireless bridge connects the gateways of two spatially separated fiber sections, each with a length of up to 20 km.

Besides electronic up-conversion, we also investigate photonic mm-wave signal generation as described in *Section 2.6.2*. In *Section 4.2*, we demonstrate a single-channel single-input single-output (SISO) mm-wave transmission at 237.5 GHz with a record data rate of 100 Gbit/s. The data are transmitted using the advanced modulation formats quadrature phase shift keying (QPSK), 8PSK, and quadrature amplitude modulation (QAM). A high quality carrier is generated with mm-wave photonic techniques by heterodyning selected lines of a mode-locked laser (MLL) on a uni-travelling carrier (UTC) photodiode (PD). After 20 m wireless transmission, the signal is down-converted to the baseband with an active mm-wave monolithic integrated circuits (MMIC) IQ mixer. In *Section 4.3*, we encode data on three adjacent MLL lines using different modulation formats and sinc-like pulse shaping (see *Section 2.1.5*). We show multi-channel SISO mm-wave transmission with a total data rate of 100 Gbit/s over distances of 10…40 m.

4.1 Multi-Gigabit Wireless Bridge Connecting Two Fiber-Optic Links

The content of this Section has been published in [J2]:

S. Koenig, J. Antes, D. Lopez-Diaz, R Schmogrow, T. Zwick, C. Koos, W. Freude, J. Leuthold, and I. Kallfass, “20 Gbit/s Wireless Bridge at 220 GHz,” *J. Opt. Commun. Netw.*, vol. 6, no. 1, pp 54-61, 2014, doi: 10.1364/JOCN.6.000054.

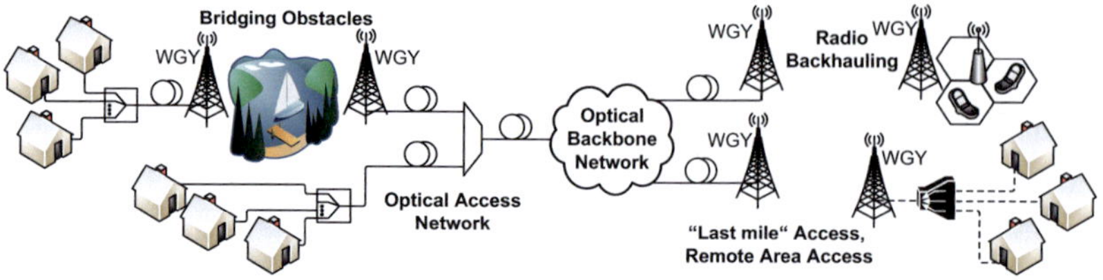

Fig. 4.1: Various applications of a high-data-rate mm-wave wireless link in a converged optical/wireless network: connecting fiber-optic data links (e.g., to bridge obstacles such as lakes, valleys, or construction sites), acting as "last mile" access (e.g., in rural and remote areas), and radio backhauling. Wireless gateways (WGYs) are the interface between the wireless world and the fiber infrastructure. In contrast with free space optical links that use visible or IR light for point-to-point communications, wireless links stay operational under adverse weather conditions such as heavy rain or fog. Reprinted from [J2] © 2014 The Optical Society.

Millimeter-wave (mm-wave) carrier frequency wireless links hold promise for supporting data rates up to 100 Gbit/s. In recent decades, both wireless and wireline data rates have been increasing exponentially [160]. In a few years, we will face wireless data rates of several tens of gigabits per second, while optical and wireless networks converge.

The large required wireless bandwidth can be provided by either free-space optical (FSO) links using infrared (IR) light or radio links operated at carrier frequencies in the mm-wave range and the lower terahertz range. FSO links with multi-gigabit capacities have already been reported [161]. However, FSO links become inoperative in adverse weather conditions such as heavy rain, fog [39], or dust [162], while mm-wave or terahertz wireless links are hampered but stay operational.

A network scenario is illustrated in Fig. 4.1. In this scenario mm-wave wireless links are used in a converged optical/wireless network whenever data transmission over optical fiber is not an available option; the wireless link may be employed to overcome natural obstacles and difficult-to-access terrain, serving as a bridging element in the fiber-optic network. The wireless link may also bring broadband services to remote areas where fiber infrastructure is not available yet. This is a flexible, cost-efficient, and easy-to-implement solution to the "last mile" access problem. Another relevant application is a high-data-rate wireless link for radio backhauling.

The interface between the wireline and wireless infrastructure is a wireless gateway (WGY) (Fig. 4.1). The WGY converts the optical data to electrical mm-wave signals, which feed a transmitting antenna. After wireless transmission, the signal is received by the antenna of another WGY, from which the signal is directed toward its final destination. This can be done by using either another wireless channel in a relay technique or a copper wire or a coaxial (TV) cable, or again employing an optical fiber.

For generating the mm-wave signal in the WGY, both *photonic* and *electronic* techniques were used in the past. So far *photonic* techniques demonstrated the encoding of on–off keying (OOK) signals onto radio-over-fiber mm waves in the 60 GHz band [45, 49, 52, 163], the 75–110 GHz band [59, 61, 65], at 120 GHz [164], and at 250 GHz [70]. Meanwhile, *electronic* upconversion techniques are also available, as metamorphic high-electron-mobility transistor (mHEMT) technology allows the implementation of active mm-wave monolithic integrated circuits (MMICs) operating beyond 200 GHz [41]. Electronic upconversion with MMIC technology could be potentially inexpensive if mass production of the MMIC is envisaged. Photonic upconversion requires optical devices such as a local oscillator (LO) laser, an optical preamplifier, and an ultrabroadband photodetector, for example, a uni-traveling carrier photodiode (UTC-PD). An important advantage of the active MMIC technology is the availability of on-chip amplifiers, which enable large power levels of > 0 dBm at the RF output port of the MMIC mixer [41]. In contrast, typical output power levels of a single UTC-PD without an electronic postamplifier are in the order of −15 to −5 dBm for mm-wave signals > 200 GHz [70, 165]. So far, at a 125 GHz carrier frequency, a 10 Gbit/s MMIC wireless link with a fiber interface has been shown [74, 166]. However, data transmission at more than 10 Gbit/s using wireless carrier frequencies beyond 200 GHz together with electronic frequency upconversion and downconversion has not yet been integrated into a fiber environment.

In this *Section* we demonstrate a wireless link at a 220 GHz carrier frequency embedded into an optical fiber environment [C18]. We compare the transmission of 16 and 20 Gbit/s non-return-to-zero (NRZ) OOK signals, and we investigate the performance of binary phase shift keying (BPSK) and quadrature phase shift keying (QPSK) signals in a bandwidth of 5 GHz employing orthogonal frequency division multiplexing (OFDM). The wireless link was connected to the fiber-optic network with opto-electronic (OE) and electro-optic (EO) converters. The RF circuits providing electronic frequency upconversion and downconversion at the WGY were implemented with MMICs in mHEMT technology [41]. In this feasibility study we used the same LO source for both the transmitter (Tx) MMIC and the receiver (Rx) MMIC to avoid the possible influences of a clock recovery device at 220 GHz. With this setup we study the influence of different fiber sections of up to 20 km before the transmitting and after the receiving WGY. We investigate different modulation depths for the OFDM signal.

4.1.1 Setup

A. Fiber-Optic Sections

The experimental setup (Fig. 4.2) consists of an active MMIC wireless link at 220 GHz connecting the WGY of two identical fiber-optic links. In each optical Tx, an external cavity laser ($\lambda = 1550$ nm) is externally modulated by a Mach–Zehnder modulator (MZM) biased at its quadrature point [intensity modulation (IM)]. The lengths L_1 and L_2 of the standard single-mode fiber (SSMF) spans after optical Tx1 and Tx2, respectively, are set to 0, 10, or 20 km. Both Rx1 and Rx2 after the respective fiber sections L_1 and L_2 comprise preamplifiers [erbium-doped fiber amplifiers (EDFAs)] with a noise figure (NF) of 5 dB. After the EDFAs, optical bandpass filters [not drawn in Fig. 4.2; bandwidth 0.6 nm (Rx1) and 2 nm (Rx2)] remove out-of-band noise. A pin-photodetector with an electrical bandwidth of 40 GHz uses direct detection (DD) for the received optical power, which is kept constant at +7 dBm. The MZMs used in this experiment are Covega Mach10 MZMs with $V_\pi = 5.5$ V and a modulation bandwidth of 13 GHz for a sensitivity drop of a 3 dB. At 20 GHz the modulation sensitivity drops by 4 dB. The MZM return loss is larger than 15 dB for frequencies below 13 GHz and decreases to minimum values of 7 dB for frequencies of 13 to 20 GHz.

B. 220 GHz Wireless Link

For brevity, we refer to the part between **A** and **B** in the experimental setup of Fig. 4.2 as the "wireless link." Starting from **A**, an RF-amplifier 1 (Amp. #1, SHF 104P, 16 dB gain, +10 dBm at 1 dB compression) boosts the received data signal to an input power level of +5 dBm, which is required for the data signal at the intermediate frequency (IF) port of the resistive mixer inside the Tx MMIC. A local oscillator at 220 GHz drives this mixer at the LO port. The upconverted data signal leaves the mixer's RF port at a power level of −20 dBm and is postamplified by a four-stage low-noise amplifier (LNA) with 20 dB small-signal gain. Thus, the total output power of the Tx MMIC module is 0 dBm. This modulated 220 GHz RF carrier feeds a lensed horn antenna with a gain of 23 dBi. The lens is an aspherical plano–convex high-density polyethylene (HDPE) lens.

After wireless transmission, an identical lensed horn antenna is used for receiving the signal. The subsequent Rx MMIC comprises the same LNA and resistive mixer stage as the Tx MMIC. The same type of mixer can be used, but this time for downconverting the modulated 220 GHz RF carrier to the baseband. The Tx and Rx MMICs are implemented in mHEMT technology featuring a 50 nm gate length, a transit frequency $f_T = 380$ GHz, a maximum oscillation frequency $f_{\max} > 600$ GHz, a transconductance of $g_{\text{m, max}} = 1800$ mS/mm, and a gain larger than 15 dB [167]. Further information about the Tx and Rx MMIC modules can be found in [168] and *Appendix A.4.1*. Two cascaded broadband RF amplifiers (Amp. #2, #3, type SHF 804TL, 21 dB gain, +15.5 dBm at 3 dB compression) amplify the baseband signal, which then is fed at point **B** to MZM 2 of the second optical Tx.

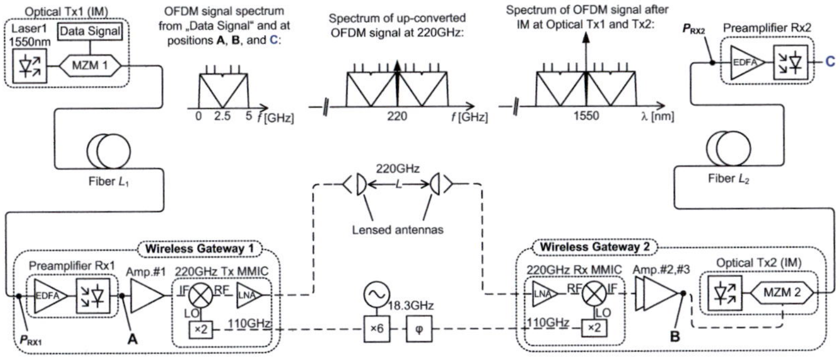

Fig. 4.2: Experimental setup of a wireless bridge operating at 220 GHz carrier frequency and connecting two fiber-optic links: optical intensity modulation (IM) of the data signals (NRZ-OOK or electrical OFDM) at the first optical transmitter (Tx) using a Mach–Zehnder modulator (MZM), transmission over a fiber section of length L_1, direct detection (DD) at wireless gateway 1 (WGY 1) using a preamplifier receiver (Rx), electronic signal upconversion to a carrier frequency of 220 GHz by the Tx mm-wave monolithic integrated circuit (MMIC), wireless transmission over an in-door test distance of $L = 0.5$ m, electronic downconversion back to the baseband by the Rx MMIC at WGY 2, and transmission over a second IM–DD link with fiber length L_2. The transmitted data signals are analyzed after the first IM–DD link (at position **A**), after wireless transmission (**B**), and after the second IM–DD link (**C**). The schematic spectra show the OFDM signal at different positions in the experimental setup. Reprinted from [J2] © 2014 The Optical Society.

In our experiments, we employ a homodyne receiver (zero-IF) architecture at the Rx MMIC, i.e., the Rx LO is phase-locked to the incoming 220 GHz RF carrier. Because we lack a clock recovery device at 220 GHz, we use the same LO source for both the Tx MMIC and the Rx MMIC. A 110 GHz RF carrier with a power of +10 dBm is input to the LO ports of both the Tx MMIC and Rx MMIC modules, where an on-chip frequency doubler converts the 110 GHz RF carrier frequency to the desired LO frequency of 220 GHz. The 110 GHz RF carrier originates from a HP W-band source module (6×18.3 GHz), and the correct phase relation between the Tx and Rx LO is manually set with a tunable phase shifter (φ). Presently, in our experimental setup, the length of the cables that distribute the 110 GHz RF carrier limits the wireless transmission distance to $L = 0.5$ m. To avoid saturating and thereby damaging the Rx MMIC during the experiments, we deliberately introduce additional losses by misaligning the HDPE lenses. It is difficult to measure the actual received RF power directly. However, by comparing the present results with the measured Rx sensitivity of the identical Rx MMIC as given in [C17], we extrapolate that the received power at the Rx MMIC is about −60 dBm for a target bit error ratio (BER) of 10^{-3}. The link budget is based on the following data: MMIC Tx output power, 0 dBm; gain of each lensed horn antenna, 23 dBi; free-space path loss, 73 dB for $L = 0.5$ m and 220 GHz (neglecting the atmospheric attenuation of 3 dB/km). Consequently, the total power budget at the receiver is −27 dBm for $L = 0.5$ m, which leaves a margin of 33 dB, corresponding to a transmission distance of more than 20 m. In [C17] we experimentally demonstrate that this power budget actually suffices to bridge a distance of 20 m for BER = 10^{-3}.

C. Transmitted OOK and OFDM Data Signals

Two different types of data signals (represented by the box "Data Signal" in Optical Tx1 of Fig. 4.2) are transmitted and then analyzed at positions **A**, **B**, and **C**.

The first kind of signal is a NRZ-OOK signal at data rates of 16 and 20 Gbit/s. The NRZ-OOK signal is generated by a MICRAM state-of-the-art high-speed digital-to-analog converter (DAC) driven by a Xilinx Virtex5 field-programmable gate array. The amplified DAC output drives the MZM 1. We measure the NRZ-OOK signal quality factor $Q = (\mu_1 - \mu_0)/(\sigma_1 + \sigma_0)$. The quantities $\mu_{1,0}$ and $\sigma_{1,0}$ denote the mean and the standard deviations of the photodetector output voltage, respectively, if ones and zeros are received. A digital communications analyzer is used to record the eye diagram of the NRZ-OOK signal and to determine the quality factor Q. The square of Q represents an electrical signal-to-noise power ratio, and for OOK signals provides an estimate of the BER: $\text{BER} \approx (1/2)\text{erfc}(Q/\sqrt{2})$. Q^2-factors corresponding to 15.6, 13.5, and 9.8 dB result in BER values of 10^{-9}, 10^{-6}, and 10^{-3}.

The second type of data is transmitted via a 5 GHz wide real-valued electrical OFDM signal. The signal is offline generated with MATLAB and physically generated by a Tektronix 7102B arbitrary waveform generator (AWG) at a sampling rate of 20 GSa/s. The size of the inverse fast Fourier transform is 256, leading to a subcarrier spacing of 78.125 MHz. The OFDM signal comprises 63 subcarriers: one unmodulated subcarrier located at the intermediate frequency of 2.5 GHz (center frequency), four BPSK-encoded pilot tones equidistantly distributed over the OFDM spectrum, and

58 subcarriers modulated with data all having an identical modulation format. We investigate an OFDM-BPSK signal and an OFDM-QPSK signal with a line rate of 4.53 and 9.06 Gbit/s, respectively. A cyclic prefix leads to an overhead of 25% and results in a reduced data rate of 3.63 and 7.25 Gbit/s, respectively. The electrical peak-to-average power ratios (PAPRs) of the OFDM-BPSK and OFDM-QPSK signals at the AWG's output are 19.8 and 24.8, respectively.

The OFDM signal spectrum is schematically depicted in Fig. 4.2 for different positions in the experimental setup. At the top, the spectrum of the real-valued electrical OFDM baseband signal as generated by the AWG and received at positions **A**, **B**, and **C** is shown. In the middle, the optical spectrum after IM at Tx1 and Tx2 is shown. At the bottom, the mm-wave signal after electronic upconversion to 220 GHz is depicted.

The OFDM signal is recorded with a Tektronix DPO70804B real-time digital oscilloscope for subsequent offline processing and signal analysis. The signal quality of the OFDM signal is evaluated in terms of the error vector magnitude (EVM). The EVM metric is well established in wireless and wireline communications. Like the Q^2-factor, the EVM can be related to the BER, see *Appendix A.3*. In our experiments, the EVM (referred to as the maximum electric field in the constellation) is calculated by averaging over 1000 OFDM symbols consisting of 58 independently modulated subcarriers each, i.e., we use a total of 58,000 pieces of complex data. For BPSK (QPSK) modulation, BERs of 10^{-9}, 10^{-6}, and 10^{-3} require EVM values of 23.4 % (16.6 %), 29.6 % (20.9 %), and 42.6 % (32.2 %), respectively.

4.1.2 Results

The NRZ-OOK and OFDM signal quality is evaluated for three cases. In *Section A*, we investigate the signal when it is transmitted only over the wireless link. In *Section B*, we evaluate the signal quality after the first OE conversion at Rx1 (position **A** for $L_1 = 0$), after the first OE conversion plus wireless transmission (position **B**), and after the first OE conversion plus wireless transmission plus the second OE conversion (at position **C** for $L_2 = 0$). Finally, in *Section C*, we include up to 20 km of SSMF both before and after the wireless transmission link and evaluate the signal quality at Rx2 (position **C**).

A. Wireless Link Only

In order to investigate the signal degradation due to the wireless link itself, we feed the signal directly from the data source (the box "Data Signal" in

Optical Tx1 of Fig. 4.2) to position **A** and analyze the signal quality after wireless transmission at position **B**. The quality of the data source is $Q^2 = 24.5$ dB and $Q^2 = 24.6$ dB for the 16 Gbit/s NRZ-OOK and the 20 Gbit/s NRZ-OOK signals, respectively. The data source's EVM for the OFDM-BPSK and OFDM-QPSK signals are 7.0 %.

The signal quality after wireless transmission is shown in Fig. 4.3, where the eye diagram and constellations are displayed. The quality factor Q^2 reduces to 17.2 dB [Fig. 4.3(a)] for 16 Gbit/s NRZ-OOK and to 16 dB [Fig. 4.3(b)] for 20 Gbit/s NRZ-OOK. These values indicate high-quality signals with a BER $< 10^{-9}$.

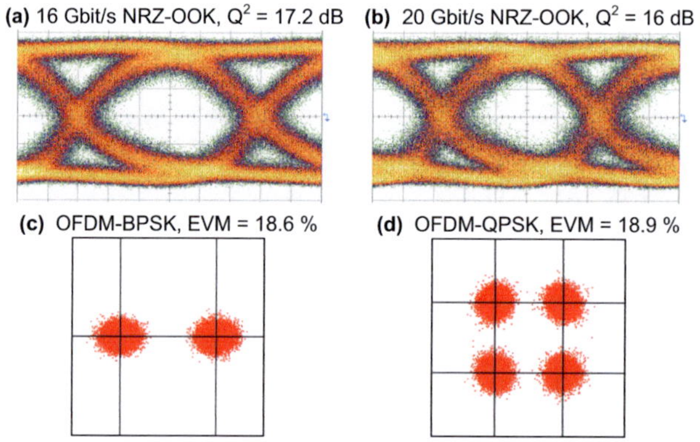

Fig. 4.3: Signal quality after transmission over the wireless link only, i.e., for data signal input at position **A** and signal analysis at position **B** : eye diagram and quality factor Q^2 for (a) 16 Gbit/s NRZ-OOK and (b) 20 Gbit/s NRZ-OOK; constellation diagram and EVM for (c) OFDM-BPSK and (d) OFDM-QPSK. For both NRZ-OOK signals and the OFDM-BPSK signal, the obtained values for Q^2 and EVM still indicate a BER $< 10^{-9}$. In the case of OFDM-QPSK, an EVM of 18.9 % corresponds to a BER$<10^{-6}$. Reprinted from [J2] © 2014 The Optical Society.

For OFDM-BPSK, the EVM rises to 18.6 % [Fig. 4.3(c)], corresponding to a BER $< 10^{-9}$, whereas for OFDM-QPSK the measured EVM of 18.9 % [Fig. 4.3(d)] corresponds to a BER $< 10^{-6}$. The RF amplifier and mixer chain clips the OFDM signal. The PAPR of the OFDM-BPSK (OFDM-QPSK) signal decreases from 19.8 (24.8) at the AWG's output at position **A** to 6.6 (6.6) after the wireless link at position **B**.

The degradation for both the NRZ-OOK and the OFDM signal is attributed to an increase of noise introduced by the wireless link.

B. Wireless Link and E/O/E Conversion Only

The wireless link is embedded between EO and OE converters. In the following subsections we investigate the quality of NRZ-OOK and OFDM, setting the fiber link lengths $L_{1,2}$ to zero.

NRZ-OOK Signals

For NRZ-OOK modulation, an electrical driver circuit (implied in the box "Data Signal" in Optical Tx1 of Fig. 4.2) amplifies the data signal to a peak-to-peak value of V_π. The modulator MZM 1 is biased at the quadrature point. A voltage deviation of $\pm V_\pi/2$ then switches the optical output intensity between maximum and zero. The signal quality is analyzed after the direct detector at the output of Rx1 (position **A** in Fig. 4.2).

Fig. 4.4(a) and Fig. 4.4(b) show the signal quality Q^2 as a function of the preamplifier input power P_{Rx1} for 16 and 20 Gbit/s NRZ-OOK, respectively. For input powers larger than −12 dBm the signal quality is high and virtually constant. Below −12 dBm, the optical signal-to-noise ratio decreases, and the receiver performance degrades. Still, the signal quality remains excellent over the whole measurement range. For an input power level of $P_{\text{Rx1}} = -12$ dBm we obtain $Q^2 \approx 22$ dB for 16 Gbit/s NRZ-OOK and $Q^2 \approx 17$ dB for 20 Gbit/s NRZ-OOK. The bandwidth limitation of MZM 1 at Tx1 is considered to be the reason for the 5 dB of Q^2 penalty between the 20 and 16 Gbit/s signals.

The signal quality at position **B** after transmission over the wireless link when varying the receiver input power P_{Rx1} is measured next. We observe a degradation of Q^2 by 6 dB (5 dB) for the 16 Gbit/s (20 Gbit/s) NRZ-OOK signal compared to the data taken at position **A**. With $Q^2 = 16$ dB, the 16 Gbit/s signal has excellent signal quality (BER $< 10^{-9}$).). The 20 Gbit/s NRZ-OOK signal shows a quality factor of $Q^2 = 12.1$ dB, which still is well above the forward error correction (FEC) limit of $Q^2 = 8.5$ dB (BER $= 4\times10^{-3}$).

The signal quality after wireless and optical transmission at position **C** is analyzed next. We keep the input power $P_{\text{Rx1}} = -12$ dBm at Rx1 constant. The input power P_{Rx2} is varied with a variable attenuator (not drawn in Fig. 4.2). Fig. 4.4 shows the corresponding quality factor Q^2 as a function of the preamplifier receiver input power P_{Rx2} (upper axes) for 16 and 20 Gbit/s NRZ-OOK [Fig. 4.4(a) and (b), respectively]. Compared to the signal quality at position **B**, no further signal degradation is seen for the 20 Gbit/s signal. For the 16 Gbit/s signal we observe an improvement of 1 dB in quality factor from $Q^2 = 16$ dB to $Q^2 = 17$ dB. This is due to the nonlinear transfer characteristic of MZM 2, where signal overshoots, undershoots, and ripples are suppressed.

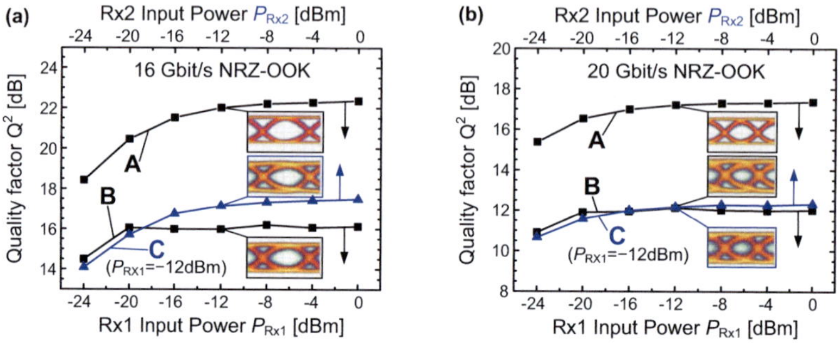

Fig. 4.4: Quality factor Q^2 as a function of receiver input powers P_{Rx1} (lower axis) and P_{Rx2} (upper axis) for (a) 16 Gbit/s NRZ-OOK and (b) 20 Gbit/s NRZ-OOK. The quality factor is measured after the first opto-electronic conversion (position **A**, $L_1 = 0$ km), after wireless transmission (position **B**), and after the second opto-electronic conversion (position **C**, $L_2 = 0$ km). For positions **A** and **B**, the lower axis denotes the Rx input power at Rx1. For position **C**, the upper axis denotes the Rx input power at Rx2. The insets show eye diagrams at the chosen operating points of $P_{\mathrm{Rx1}} = P_{\mathrm{Rx2}} = -12$ dBm. With $Q^2 = 17$ dB at position **C**, the 16 Gbit/s signal has excellent signal quality (BER < 10^{-9}). The 20 Gbit/s NRZ-OOK signal shows a quality factor of $Q^2 = 12.1$ dB, which is still above the forward error correction limit of $Q^2 = 8.5$ dB (BER = 4×10^{-3}). Reprinted from [J2] © 2014 The Optical Society.

OFDM Signals

The OFDM transmitter comprises an electrical driver amplifier to boost the OFDM signal before MZM 1. The setup is identical to the OOK setup as described previously. After the driver amplifier at the MZM RF port, the root mean square (rms) value of the OFDM signal is $V_{\mathrm{rms}} \approx 0.4$ V and the PAPRs of the OFDM-BPSK and OFDM-QPSK signals are 12.1 and 13.4, respectively.

We define the optical modulation depth as $m = \sqrt{2}\, V_{\mathrm{rms}} / V_\pi$. With $V_\pi = 5.5$ V for the MZM, a modulation depth of $m_1 = 10$ % around the quadrature point results after Tx1. In contrast to the 20 Gbit/s OOK signal, the MZM offers sufficient bandwidth for the 5 GHz wide OFDM signal. However, the nonlinear sinusoidal transfer function of the MZM impairs the OFDM signal. At Rx1, we choose an operating point at $P_{\mathrm{Rx1}} = -12$ dBm. The modulation depth of $m_1 = 10$ % ensures that after Amp #1 the required input power level of +5 dBm is obtained at the IF port of the Tx MMIC.

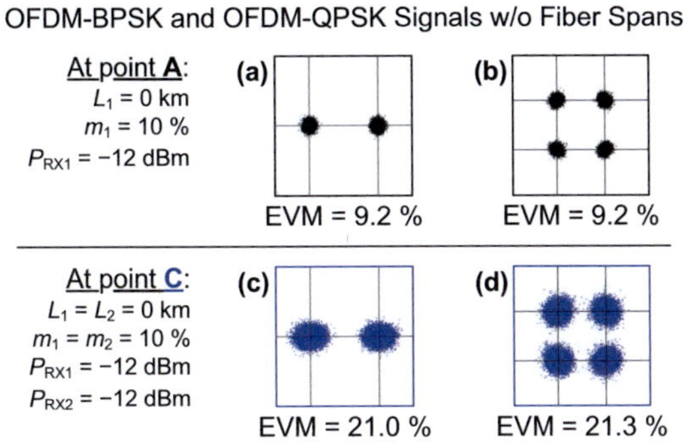

Fig. 4.5: Constellation diagrams and EVM for (a), (c) OFDM-BPSK and (b), (d) OFDM-QPSK after the first opto-electronic conversion (position **A**, $L_1 = 0$ km) and after the second opto-electronic conversion including the wireless link (position **C**, $L_2 = 0$ km), respectively. The modulation depth and receiver input power for both optical links was $m_1 = m_2 = 10$ % and $P_{\text{Rx1}} = P_{\text{Rx2}} = -12$ dBm, respectively. At position **C** we obtain EVM = 21.0 % for OFDM-BPSK and EVM = 21.3 % for ODFM-QPSK, which is still below the EVM limits for a BER of 10^{-9} and 10^{-5}, respectively. Reprinted from [J2] © 2014 OSA.

Fig. 4.5(a) and Fig. 4.5(b) show a constellation diagram overlay of all 58 independently modulated subcarriers after Rx1 (position **A**) for the OFDM-BPSK and OFDM-QPSK signals, respectively. For both signals, the EVM has increased by 2.2 percentage points to 9.2 % compared to the signal directly measured from the data source. For both BPSK and QPSK modulation, this is still an excellent signal quality. The PAPR of the OFDM-BPSK (OFDM-QPSK) signal after Rx1 is 10.9 (11.5).

Fig. 4.5(c) and Fig. 4.5(d) show the constellation diagrams at position **C**, i.e., the system including EOE conversion and the wireless link, but with $L_{1,2} = 0$ km. The optical modulation depth m_2 for the second IM–DD link is adjusted to $m_2 = m_1 = 10$ %. As with Rx1, we choose $P_{\text{Rx2}} = -12$ dBm as an operating point for Rx2. Due to the signal clipping by Amp #3, the PAPR of the OFDM-BPSK (OFDM-QPSK) signal after Rx2 is reduced to 6.6 (7.0). At position **C** we obtain EVM = 21.0 % for OFDM-BPSK and EVM = 21.3% for ODFM-QPSK, which is still below the EVM limits for a BER of 10^{-9} and 10^{-5}, respectively.

C. Wireless Link, E/O/E Conversion & Fiber Links

In the following *Subsections* we investigate the quality of NRZ-OOK and OFDM signals including variable fiber links with lengths $L_{1,2}$ from 10 km to 20 km.

NRZ-OOK Signals

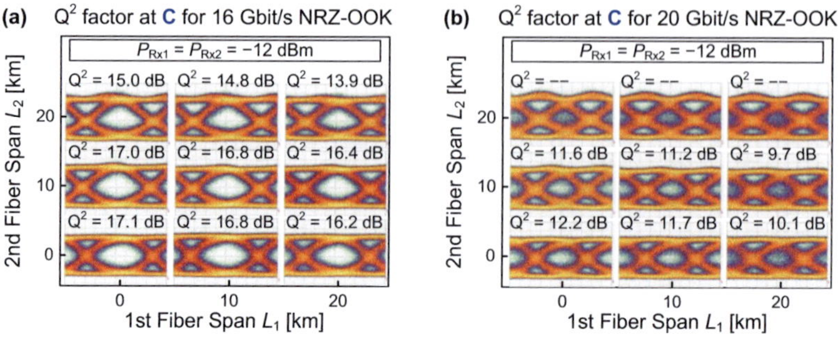

Fig. 4.6: Eye diagram and quality factor Q^2 as a function of the fiber lengths L_1 and L_2 before and after the wireless link, respectively, for (a) 16 Gbit/s NRZ-OOK and (b) 20 Gbit/s NRZ-OOK. Q^2 decreases (eye is closing) for increasing fiber span lengths, due to fiber dispersion. For the 16 Gbit/s NRZ-OOK signal, Q^2 =13.9 dB (BER $\approx 10^{-6}$) is obtained after a total fiber span of 40 km ($L_1 = L_2 = 20$ km). The quality of the 20 Gbit/s NRZ-OOK signal without fiber transmission is already impaired by the bandwidth limitation of Tx1 and the MMIC modules ($Q^2 = 12.2$ dB for $L_1 = L_2 = 0$ km). Fiber dispersion further reduces the quality factor to $Q^2 = 9.7$ dB (corresponding to a BER of 10^{-3}), which was the lowest measurable value for the 20 Gbit/s NRZ-OOK, for a total fiber length of 30 km. Reprinted from [J2] © 2014 The Optical Society.

Fig. 4.6 shows the measured eye diagrams and quality factors for 16 and 20 Gbit/s NRZ-OOK signals [Fig. 4.6(a) and Fig. 4.6(b), respectively] at position **C** for different combinations of fiber length L_1 (horizontal axis) before and L_2 (vertical axis) after the wireless link. Note that the eye diagrams shown for $L_1 = L_2 = 0$ km correspond to the results discussed in *Subsection B*. It can be observed that the Q^2-factor decreases for increasing fiber spans. Since only fiber has been added to the system, the observed degradation of Q^2 is attributed to fiber dispersion and not to noise. In the case of the 16 Gbit/s NRZ-OOK signal, a Q^2-factor of 13.9 dB is measured after a total fiber span of 40 km ($L_1 = L_2 = 0$ km), which still corresponds to a BER of about 10^{-6}. As discussed in *Subsection B*, the quality of the 20 Gbit/s NRZ-OOK signal without fiber transmission is already impaired by the bandwidth limitation of Tx1 (MZM 1) and the MMIC modules ($Q^2 = 12.2$ dB for $L_1 = L_2 = 0$ km). Moreover, for 20 Gbit/s NRZ-OOK, the signal impairment due to fiber dispersion is stronger than for 16 Gbit/s NRZ-OOK. Figure 6(b) shows that $Q^2 = 9.7$ dB (corresponding to a BER of 10^{-3}) was the lowest measurable value for 20 Gbit/s NRZ-OOK for a total fiber length of 30 km. For $L_2 = 20$ km, the eye diagrams are closing due to chromatic dispersion, and no Q^2-factor is measurable anymore.

The operation point of receiver Rx1 after fiber span 1 and the operation point of receiver Rx2 after fiber span 2 were chosen to be $P_{Rx1} = P_{Rx2} = -12$ dBm for all transmission experiments. This operation point was chosen based on the results given in Fig. 4.4.

OFDM Signals

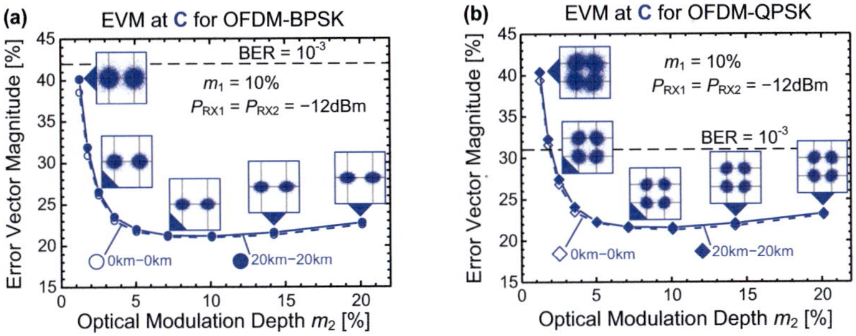

Fig. 4.7: EVM as a function of the optical modulation depth m_2 on the second fiber span for (a) OFDM-BPSK and (b) OFDM-QPSK. Virtually no EVM penalty is observed after signal transmission over 40 km ($L_1 = L_2 = 20$ km) compared to the case without fiber transmission ($L_1 = L_2 = 0$ km). Fiber dispersion does not impair the OFDM signal quality. For low modulation depths, the signal is degraded by noise. At higher modulation depths, the signal quality is mainly distorted by nonlinearities causing intermodulation among the OFDM subcarriers. The insets show the constellation diagrams at the respective modulation depths. Reprinted from [J2] © 2014 The Optical Society.

We also investigate the EVM of the OFDM-BPSK and OFDM-QPSK signals after Rx2 at position **C** as a function of the optical modulation depth m_2 after Tx2 and as a function of different fiber lengths L_1 and L_2. The modulation depth m_1 on the first optical link is kept constant at $m_1 = 10$ %. Possibly, reducing this modulation depth would have led to even better results, but the electrical mixer behind Rx1 requires a minimum input power. Fig. 4.7 shows that a minimum EVM is found for $m_2 \approx 7$ % irrespective of the fiber length and modulation format. For lower modulation depths, the signal is impaired by noise. At higher modulation depths, the signal quality is mainly degraded by nonlinearities causing intermodulation among the OFDM subcarriers. Signal fading due to chromatic dispersion does not affect the OFDM signal, since the first complete extinction of the highest subcarrier frequency of 5 GHz would occur only after about 150 km [110]. Therefore, we conclude that for fiber lengths L_1 and L_2 of more than 20 km the OFDM signal BER remains better than 10^{-3}. Consequently, error-free transmission is possible when employing second-generation FEC.

4.1.3 Conclusions

Multi-gigabit wireless links based on electronic upconversion and downconversion techniques using MMIC technology and operating at carrier frequencies beyond 200 GHz have a great potential for being integrated transparently into fiber-optic communication networks. In this paper, data transmission of both a NRZ-OOK signal with up to 20 Gbit/s and a 9 Gbit/s OFDM signal was successfully demonstrated over a wireless bridge. MMIC technology could also be used to lock the local oscillator at the receiver to the mm-wave carrier frequency. The wireless bridge is operated at a carrier frequency of 220 GHz and connects two fiber-optic links, each with a span of up to 20 km.

4.2 Single-Channel 100 Gbit/s Wireless Link with mm-Wave Photonics

The content of this Section has been published in the conference paper [C10]. Additional results for circular 8QAM were added in Fig. 4.9(b) and Fig. 4.10(b).

S. Koenig, F. Boes, D. Lopez-Diaz, J. Antes, R. Henneberger, R. M. Schmogrow, D. Hillerkuss, R. Palmer, T. Zwick, C. Koos, W. Freude, O. Ambacher, I. Kallfass, and J. Leuthold, "100 Gbit/s Wireless Link with mm-Wave Photonics," *Optical Fiber Communication Conference (OFC)*, Anaheim (CA), USA, March 2013, postdeadline paper PDP5B.4., doi: 10.1364/NFOEC.2013.PDP5B.4.

At home or on the move – communication technology has become an essential part of our daily lives. In addition to well-established applications such as surfing the internet, emailing, online shopping, or social networking, a wealth of upcoming services like high definition (3D) video and cloud computing will significantly increase the amount of both wireline and wireless data traffic. For the customer experience it is essential to provide high-speed omnipresent internet access. Fiber-optic wireline communications offers aggregate data rates of several Tbit/s. However, sometimes an end-to-end fiber connection is absent (see *Section 1.4*), e.g., in difficult-to-access terrains and certain rural areas, or an already existing fiber connection fails, because it has been damaged during construction works or during a natural disaster. In such cases, a wireless link with a data rate > 100 Gbit/s could still provide the customer with transparent high-speed data access, even under difficult weather conditions, if the carrier frequency is chosen properly.

The broad atmospheric mm-wave transmission window from 200 GHz to 300 GHz offers the bandwidth for a 100 Gbit/s wireless link, while free-space optical links become inoperative in adverse weather conditions. In addition, the channel capacity can be increased with spectrally efficient in-phase/quadrature (I/Q) modulation formats. A direct and transparent transition from fiber to wireless is possible by photonic mm-wave generation using optical heterodyning on a photodiode (PD). Even polarization-multiplexed signals in a fiber can be transmitted at the expense of two wireless 2×2 multiple-input multiple-output (MIMO) transmission channels. The polarization demultiplexing can be done by digital signal processing (DSP) at the receiver (Rx).

Previously, optical heterodyning was used to generate multi-gigabit wireless signals in the 60 GHz band [45], the 75 GHz to 110 GHz band (W-band) [67-69], at 120 GHz [164], and up to 300 GHz [73]. A line rate of 108 Gbit/s was reported for a fiber-wireless system [69], where I/Q data were transmitted at 100 GHz carrier frequency over a 1 m 2×2 MIMO wireless link. Beyond 110 GHz, simple on-off keying modulation and envelope detection with Schottky barrier diodes at the Rx represents the state of the art, and data rates up to 28 Gbit/s have been shown [73].

In this *Section*, we generate a 100 Gbit/s photonic mm-wave carrier at a frequency of 237.5 GHz by heterodyning two subcarriers of a mode-locked laser (MLL) on a uni-travelling-carrier (UTC) PD [165]. Since the MLL lines are frequency locked, the generated carrier is exceptionally pure and stable. We transmit 50 Gbit/s quadrature phase shift keying (QPSK), 75 Gbit/s 8PSK, and quadrature amplitude modulation (QAM) with data rates of 50 Gbit/s and 100 Gbit/s for 16QAM. The free-space transmission distance in all cases is 20 m. At the Rx we implement a single-chip I/Q down-converter [169] with an active mm-wave monolithic integrated circuit (MMIC) that directly down-converts the 100 Gbit/s signal from 237.5 GHz to the baseband, see *Section 2.6.1*. To the best of our knowledge, this is the first demonstration of a 100 Gbit/s single-input single-output (SISO) wireless link. With 2×2 MIMO wireless transmission [67, 69], a 200 Gbit/s link can be established.

4.2.1 Setup

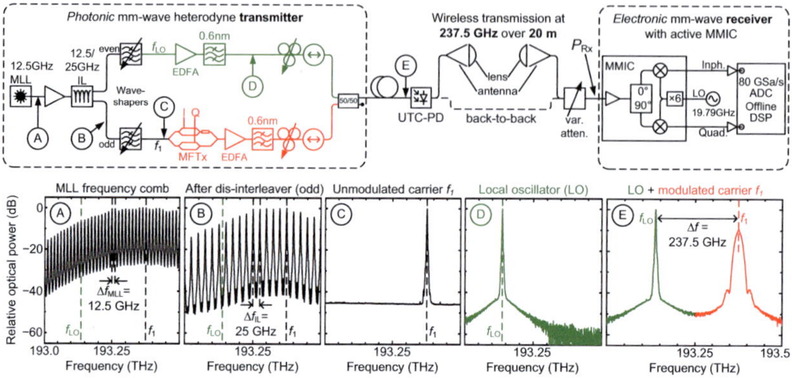

Fig. 4.8: Experimental setup for the 100 Gbit/s wireless link at 237.5 GHz over 20 m. The mm-wave signal is generated photonically: A multi format transmitter (MFTx) encodes I/Q-data onto a single subcarrier f_1 of a mode-locked laser (MLL). A second MLL subcarrier f_{LO} serves as local oscillator (LO) in the heterodyning process on a uni-travelling-carrier photodiode (UTC-PD). The insets show spectra at positions A…E from 193 THz to 193.5 THz. At the receiver, the mm-wave signal is directly down-converted to baseband using active mm-wave monolithic integrated circuits (MMIC). Analog-to-digital converters (ADC) digitize the I- and Q-components for digital signal processing (DSP). Reprinted from [C10] © 2013 The Optical Society.

Fig. 4.8 shows the experimental setup with mm-wave transmitter (Tx) and Rx. The MLL (Time Bandwidth Products ERGO XG) in the transmitter outputs a frequency comb with subcarrier spacing $\Delta f_{MLL} = 12.5$ GHz. The linewidth of a single subcarrier is about 1 kHz [170]. Inset A shows the relative MLL power spectrum. The vertical dashed lines indicate the subcarriers that will be used as local oscillator (LO, $f_{LO} = 193.138$ THz $\triangleq$ $\lambda_{LO} = 1552.22$ nm) and optical data carrier $f_1 = 193.3755$ THz. The MLL comb is amplified by an erbium doped fiber amplifier (EDFA) and passed through a 12.5/25 GHz dis-interleaver (IL) that separates the comb into even and odd subcarriers. Inset B shows the spectrum at the odd output port for f_1 (the even output port for f_{LO} is not shown). After the IL, the MLL subcarriers adjacent to f_1 and f_{LO}, respectively, are suppressed by 35 dB. This facilitates filtering of the two carriers f_1 and f_{LO} using two waveshapers (Finisar 4000E). Subcarrier f_1 (inset C) is then modulated by a multi-format (MF) Tx [83] with I/Q data ($2^{15}-1$ pseudorandom bit sequence, non-return-to-zero, 12.5 GBd and 25 GBd). The LO f_{LO} (inset D) and the data carrier f_1 are each amplified by an EDFA (noise figure NF = 3.5 dB), and filtered with a

0.6 nm band-pass filter. For our experiment, we aligned LO and data carrier in polarization and combined them by a polarization maintaining 3 dB coupler (inset E). As mentioned previously, even polarization-multiplexed optical signals can be transmitted at the expense of two wireless 2×2 MIMO channels. Polarization demultiplexing takes place at the Rx. The combined signal (total power +14 dBm) is superimposed on an ultra-fast, linear UTC-PD photomixer similar to the one shown in [165], *Appendix A.5.2*. The PD current (6.5 mA average value) comprises the mm-wave signal at $f_1 - f_{LO} = 237.5$ GHz. Around this frequency our MMIC Rx works best. We measured a total mm-wave signal power of −13.5 dBm using an Erickson PM4 calorimeter.

The modulated mm-wave signal is radiated by a cylindrical horn antenna and focused by a plano-convex dielectric lens. After 20 m wireless transmission (≈ 106 dB free-space path loss) an identical lens-antenna combination collects the signal for the Rx. The combined Tx-Rx antenna gain is 86 dBi. After the Rx antenna, the Rx input power P_{RX} to the MMIC can be adjusted with a variable attenuator.

The broadband active MMIC I/Q Rx chip (conversion gain ≈ 4 dB at 240 GHz) is designed for the frequency range 200 GHz to 280 GHz [169], see *Appendix A.4.2*. It is implemented in a 35 nm mHEMT technology. A three stage low-noise amplifier (LNA, 30 dB gain, −30 dBm saturation input power, NF = 9 dB) amplifies the received mm-wave signal. The subsequent Lange coupler splits the mm-wave signal and provides a 90 degree phase difference between its two output ports. The split signals are directly down-converted to the baseband by two subharmonic resistive mixers. In the subharmonic approach, a strong LO is provided at half the mm-wave signal frequency (118.75 GHz), and mixing is done with the LO's second harmonic. In our setup, the LO is generated from an external 19.79 GHz oscillator, the frequency of which is sextupled on-chip. In the baseband, the signal's I/Q components are analog-to-digital converted (ADC) by an 80 GSa/s real-time sampling scope for further digital signal processing including channel equalization, potentially polarization demultiplexing, carrier recovery, and demodulation.

4.2.2 Results

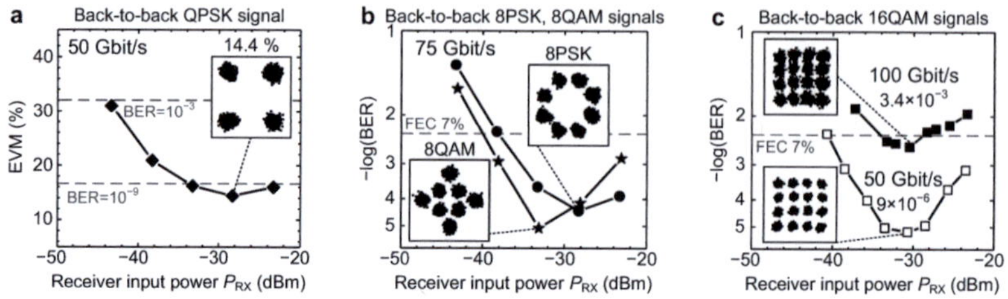

Fig. 4.9: Back-to-back measurement results: EVM for (a) 50 Gbit/s QPSK (◆) and BER for (b) 75 Gbit/s 8PSK (●) and circular 8QAM (★), and (c) 50 Gbit/s (□) and 100 Gbit/s (■) 16QAM. The insets show the constellation diagrams at the best measurement points. Modified from [C10] © 2013 OSA.

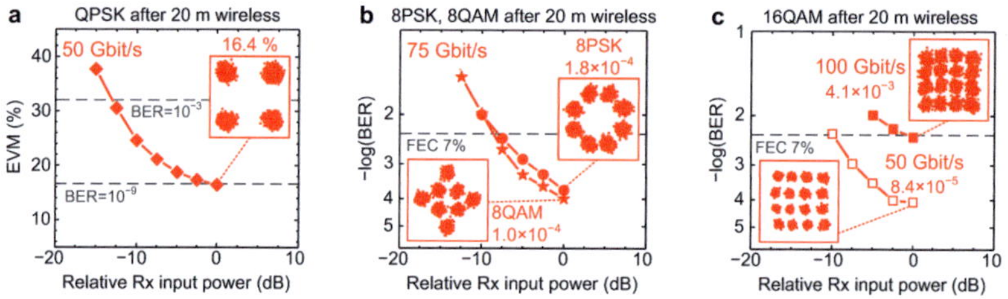

Fig. 4.10: Wireless transmission results after 20 m for (a) 50 Gbit/s QPSK (◆), (b) 75 Gbit/s 8PSK (●) and circular 8QAM (★), and (c) 50 Gbit/s (□) and 100 Gbit/s (■) 16QAM. The receiver input power is related to the same (but unknown) reference power for all graphs. A relative Rx power of −10 dB corresponds to an Rx attenuation of 10 dB. Virtually no signal impairment is seen compared to the back-to-back case. Modified from [C10] © 2013 OSA.

First we measure the back-to-back (B2B) signal quality as a function of the Rx input power P_{Rx}. Refer to *Appendix A.6* for details on the B2B setup. To this end, we directly connect the UTC-PD photomixer module to the variable attenuator at the Rx. Fig. 4.9(a) shows the measured error vector magnitude (EVM) for the 50 Gbit/s QPSK signal (◆). In this case, we specify EVM values instead of bit error ratios (BER), since the signal quality is so good that too few errors are counted within our recording length of 2×10^6 symbols. The horizontal dashed lines indicate the corresponding calculated BER values (see *Appendix A.3*) of 10^{-3} and 10^{-9}. With increasing P_{Rx}, the EVM decreases (improves) up to a point ($P_{\mathrm{Rx}} \approx -30$ dBm) where the so far dominating Rx noise becomes comparable to the Tx noise. For larger P_{Rx}, the signal is distorted because the LNA starts to saturate, and its noise figure increases.

For $P_{Rx} = -28$ dBm, we obtain an EVM of 14.4 % which indicates a BER < 10^{-9}. The inset shows the corresponding optimum constellation diagram.

Fig. 4.9(b) shows the measured B2B BER curves for 75 Gbit/s 8PSK (●). At $P_{Rx} = -28$ dBm, we measure the best BER value of 3.9×10^{-5}, which is clearly below the limit of 4.5×10^{-3} for forward error correction (FEC) with 7 % overhead [171] as indicated by the horizontal dashed line. Fig. 4.9(b) also shows the B2B measured BER curves for 75 Gbit/s circular 8QAM. At $P_{Rx} = -33$ dBm, we measure the best BER value of 8.1×10^{-6}, which is clearly below the 7 % FEC limit. At higher Rx input powers, the signal is affected by Rx nonlinearities. For 8QAM, a 2…3 dB advantage in terms of receiver input power is observed compared to 8PSK. As already pointed out in *Section 2.1.3*, the circular 8QAM constellation is known to be the best eight-point constellation.

Fig. 4.9(c) gives the B2B results for the 50 Gbit/s (□) and 100 Gbit/s (■) 16QAM signals. At $P_{Rx} \approx -30$ dBm, we achieve a BER of 9×10^{-6} for the 50 Gbit/s signal and a BER of 3×10^{-3} for the 100 Gbit/s signal. For both signals this is well below the 7 % FEC limit. At high Rx input powers, the signal is affected by Rx nonlinearities.

Fig. 4.10 shows the results for wireless transmission over a distance of 20 m. Since mm-wave power measurements below 1 μW (−30 dBm) become unreliable with our calorimetric power meter, we plot EVM and BER as a function of the relative power determined by the Rx attenuator setting: A relative Rx power of −10 dB corresponds to an Rx attenuation of 10 dB. Virtually the same behavior as in the B2B case is observed. All signals stay below the 7 % FEC threshold. Obviously, the 20 m transmission distance does not impair the signal significantly. The curves suggest that we hit the optimum Rx input power of −30 dBm for an attenuation of 0 dB.

4.2.3 Conclusions

We demonstrated SISO mm-wave transmission at 237.5 GHz with a record data rate of 100 Gbit/s.

The high quality carrier was generated with mm-wave photonic techniques by heterodyning selected lines of a MLL on a UTC-PD. After 20 m wireless transmission, the signal was down-converted to the baseband with an active MMIC. MIMO transmission and larger Tx power could improve the transmission to 200 Gbit/s over hundreds of meters.

4.3 Multi-Channel 100 Gbit/s Wireless Link with mm-Wave Photonics

The content of this Section has been published in the journal publication [J3] and in its supplementary information:

S. Koenig, D. Lopez-Diaz, J. Antes, F. Boes, R. Henneberger, A. Leuther, A. Tessmann, R. Schmogrow, D. Hillerkuss, R. Palmer, T. Zwick, C. Koos, W. Freude, O. Ambacher, J. Leuthold, and I. Kallfass, "Wireless sub-THz communication system with high data rate," *Nature Photon.*, vol. 7, no. 12, pp. 977-981, 2013, published online 13 October 2013, doi: 10.1038/nphoton.2013.275.

In communications, the frequency range 0.1–30 THz is essentially terra incognita. Recently, research has focused on this terahertz gap, because the high carrier frequencies promise unprecedented channel capacities [172]. Indeed, data rates of 100 Gbit/s were predicted [173] for 2015. We present, for the first time, a single-input and single-output wireless communication system at 237.5 GHz for transmitting data over 20 m at a data rate of 100 Gbit/s. This breakthrough results from combining terahertz photonics and electronics, whereby a narrow-band terahertz carrier is photonically generated by mixing comb lines of a mode-locked laser in a uni-travelling-carrier photodiode. The uni-travelling-carrier photodiode output is then radiated over a beam-focusing antenna. The signal is received by a millimeter-wave monolithic integrated circuit comprising novel terahertz mixers and amplifiers. We believe that this approach provides a path to scale wireless communications to Tbit/s rates over distances of > 1 km.

Data rates in both fiber-optic and wireless communications have been increasing exponentially over recent decades. For the upcoming decade this trend seems to be unbroken, at least as far as fiber-optic communications is concerned. In wireless communications, however, the spectral resources are extremely limited because of the heavy use of today's conventional frequency range up to 60 GHz. Even with spectrally highly efficient quadrature amplitude modulation (QAM) and the spatial diversity achieved with multiple-input and multiple-output (MIMO) technology, a significant capacity enhancement to multi-gigabit or even terabit wireless transmission requires larger bandwidths, which are only available in the high millimeter-wave and terahertz region. Between 200 and 300 GHz there is a transmission window with low atmospheric losses [36]. In contrast to free-space optical links, millimeter-wave or terahertz transmission is less affected by adverse weather conditions like rain and fog [39, 40].

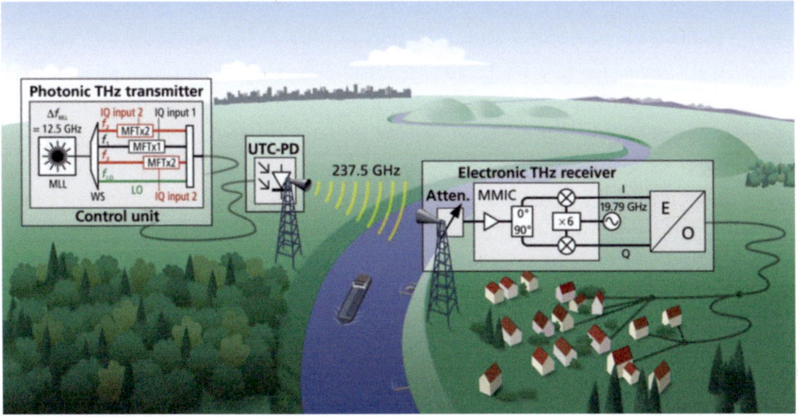

Fig. 4.11: Prospective application scenario for a long-range, high-capacity wireless communication link at THz frequencies. Here, the wireless link bridges a broad river in difficult-to-access terrain to provide high-speed internet access in remote and rural areas. The figure shows key elements of our 100 Gbit/s SISO wireless link at 237.5 GHz carrier frequency. The wireless signal is generated photonically: a mode-locked laser in the control unit outputs a frequency comb with frequency separation $\Delta f_{\mathrm{MLL}} = 12.5$ GHz. Three adjacent comb lines f_1, f_2, and f_3 and a line denoted LO are selected by a waveshaper (WS). Two multi-format transmitters (MFTx1,2) encode different in-phase (I) and quadrature (Q) data onto the central carrier f1 and the group of two carriers f_2 and f_3. The modulated data carriers f_1, f_2, and f_3 and the unmodulated LO f_{LO} are mixed by a remote UTC-PD. Three distinct millimeter-wave carriers encoded with data at $f_2 - f_{\mathrm{LO}} = 225$ GHz, $f_1 - f_{\mathrm{LO}} = 237.5$ GHz, and $f_3 - f_{\mathrm{LO}} = 250$ GHz are then radiated over a beam-focusing antenna, and received by a second beam-focusing antenna (Atten. = variable attenuator). An electronic receiver comprising active MMICs acts as an optical transmitter via an electro-optical converter (E/O) for transporting the data over a connecting fiber to any other optical receiver. Reprinted from [J3] © 2013 NPG.

Here, we present for the first time a single-input single-output (SISO) wireless 100 Gbit/s link with a carrier frequency of 237.5 GHz. By combining state-of-the-art terahertz photonics and electronics and by utilizing the large frequency range in the terahertz window between 200 and 300 GHz, we realize a wireless 100 Gbit/s link with SISO technology, that is, a link with one transmit antenna and one receive antenna. To date, 100 Gbit/s wireless links have only been demonstrated at lower carrier frequencies around 100 GHz [67-69] over a wireless distance of ~1 m, with a bit error ratio (BER) of $\sim 1 \times 10^{-3}$. Because of the limited bandwidth, these systems relied on optical polarization multiplexing and spatial MIMO with more than one wireless transmitter and receiver. Here, a 100 Gbit/s wireless transmission capacity is achieved without resorting to MIMO technology.

We envisage various applications [172, 174, 175] for such a high-capacity wireless link (Fig. 4.11). If an end-to-end fiber connection is absent and the deployment of a new fiber link is not economical, as might be the case in difficult-to-access terrains and certain rural areas (last mile problem), or if an already existing fiber connection fails, a permanent or *ad hoc* wireless connection could help. Furthermore, we anticipate indoor applications, such as high-speed wireless data transfers between mobile terminals and desktop computers. Fig. 4.11 presents a schematic of our 100 Gbit/s wireless experiment embedded into an application scenario where an obstacle, here a broad river, is bridged by the wireless link. We first discuss the general system concept, and then provide further details.

4.3.1 Setup

For the transmitter (Tx) we use a terahertz photonics technology set-up (Fig. 4.11). We generate exceptionally pure and stable terahertz carriers by heterodyning frequency-locked laser lines11. A control unit contains a single mode-locked laser (MLL), selects the appropriate frequency-locked comb lines, and modulates data on the carrier lines. An optical fiber transmits the modulated carriers together with an unmodulated comb line, which acts as a remote local oscillator (LO), to a remote uni-travelling-carrier photodiode (UTC-PD). By photomixing the LO and the modulated carriers, radiofrequency signals are generated. Optical heterodyning has already been used in earlier works to implement multi-gigabit wireless systems in the 60 GHz band [45, 52, 163] in the W-band (75–110 GHz [52, 61, 67, 68, 176]), at 120 GHz [72, 164] and at carrier frequencies beyond 200 GHz [70, 73].

For the electronic in-phase/quadrature (IQ) receiver (Rx), we use a custom-developed, active millimeter-wave monolithic integrated circuit (MMIC) with a radiofrequency bandwidth of 35 GHz [169], [C6]. This is, to the best of our knowledge, the first active broadband IQ mixer at 237.5 GHz. The Rx comprises a low-noise amplifier (LNA) and a subharmonic downconversion IQ mixer, and is realized in a metamorphic high electron mobility transistor (mHEMT) technology (see *Appendix A.4.2*) featuring a gate length of 35 nm and a cutoff frequency of more than 900 GHz [167, 177]. The complex data are directly downconverted to the baseband and separated into I and Q signals. Previous works [52, 67, 68] in the W-band have illustrated the importance of a high carrier frequency. However, to date, no direct downconversion to baseband has been used due to a lack of IQ mixers covering the full W-band. For carrier frequencies beyond 110 GHz [70, 72, 73, 164], simple on–off keying modulation and envelope detection with Schottky barrier diodes have comprised the state of the art, and data rates up to 28 Gbit/s over 0.5 m have been demonstrated [73].

In our photonic terahertz Tx, a single MLL (Time Bandwidth Products ERGO XG) outputs a frequency comb with frequency separation $\Delta f_{\mathrm{MLL}} = 12.5$ GHz. A programmable optical filter (Finisar waveshaper 4000E) selects the desired optical carriers from the comb: the LO at $f_{\mathrm{LO}} = 193.138$ THz ($\lambda_{\mathrm{LO}} = 1552.22$ nm), a central carrier at $f_1 = 193.3755$ THz ($\lambda_1 = 1550.31$ nm) and two adjacent carriers at $f_{2,3} = f_1 \mp \Delta f_{\mathrm{MLL}}$. The central carrier and the group of two adjacent carriers are each modulated with different IQ data, provided by two independent multi-format transmitters [83] MFTx 1 and MFTx 2. The modulated carriers and the LO are amplified, aligned in polarization, combined, and superimposed on a UTC-PD (see *Appendix A.5* for UTC-PD details) similar to the one shown in ref. [165] for photomixing. The photodiode current comprises signals at the three intermediate frequencies $f_{1,2,3} - f_{\mathrm{LO}}$ that are radiated from a horn antenna and focused by an aspherical plano-convex lens.

The UTC-PD current also comprises unwanted intercarrier mixing products at $(f_3 - f_1) = (f_1 - f_2) = 12.5$ GHz and $(f_3 - f_2) = 25$ GHz, which are filtered by the rectangular waveguide (TE10 mode cutoff frequency $f_c = 174$ GHz) connecting the UTC-PD to the antenna. The UTC-PD is operated with a total optical input power of +14 dBm, which leads to an average photodiode current of 6.5 mA and an output power of −13.5 dBm for a single modulated terahertz channel at $f_1 - f_{\mathrm{LO}} = 237.5$ GHz. We estimate that the single sideband (SSB) phase noise of the optically generated 237.5 GHz carrier is −33.1, −66.4, −80.8 and −81.9 dBc/Hz at offset frequencies of 100 Hz, 1 kHz, 10 kHz and 100 kHz, respectively (see *Appendix A.6.2*).

The receiver is equipped with a similar lens-and-horn antenna set-up, and is positioned at distances $d =$ 5, 10, 20 and 40 m. The combined transmitter–receiver lens-and-horn antenna gain is 86 dBi (compared to an isotropic antenna). The beamwidth is < 2°, which suits wireless point-to-point connections. The small wavelength of ~1 mm would allow the realization of very compact antenna arrays. For $f = 237.5$ GHz, $d = 10$ m and vacuum speed of light c, the free-space path loss is $L_{\mathrm{FSPL}} = (4\pi d f / c)^2$, corresponding to 100 dB.

After the Rx antenna, a variable waveguide attenuator allows the input power to be adjusted for our MMIC Rx. Owing to the integrated LNA stage, the Rx ideally operates at an input power of only −32 dBm. In the MMIC

Rx, the LNA stage provides ~30 dB gain and forwards the signal to an IQ mixer. The subharmonic mixer operates with a LO with half the frequency (118.75 GHz) of the transmitted center carrier. This LO signal is derived via a frequency multiplier chain from a synthesizer operating at 19.79 GHz. The estimated SSB phase noise of the Rx LO at 237.5 GHz is −44.4, −56.4, −56.4 and −78.4 dBc/Hz at offset frequencies of 100 Hz, 1 kHz, 10 kHz and 100 kHz, respectively. The complete IQ Rx is integrated on a chip area of 2.5×1 mm^2 and packaged into a split-block waveguide module. The packaged Rx module features a measured conversion gain of 3.8 dB and a noise figure of < 10 dB [169], [C6]. The downconverted I and Q baseband signals are digitized by an 80 GSa/s analogue-to-digital converter (Agilent real-time oscilloscope DSO-X-93204A). Subsequent offline processing 1[178, 179] includes channel equalization, carrier recovery, filtering and signal demodulation. See *Appendix A.6* for more details on the setup.

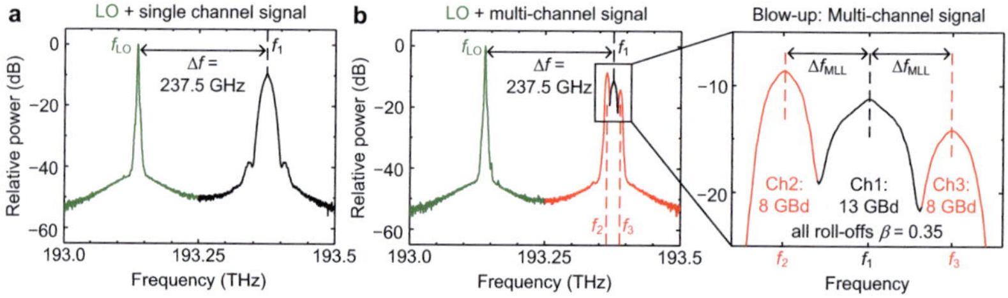

Fig. 4.12: Optical spectra at uni-travelling carrier photodiode (UTC-PD) for photonic THz signal generation. (a), Single-channel signal configuration. The frequency spacing Δf between the unmodulated optical local oscillator f_{LO} and the modulated data carrier f_1 (12.5 GBd 16QAM or 25 GBd QPSK, 8QAM or 16QAM) defines the generated wireless carrier frequency after optical heterodyning by the UTC-PD. (b), Multiple-channel signal configuration. Three data carriers with channel spacing Δf_{MLL} = 12.5 GHz are modulated with 13 GBd 16QAM (Ch1) and 8 GBd 8QAM (Ch2 and Ch3). Sinc-like pulses with a raised-cosine spectrum and a roll-off factor β = 0.35 are transmitted in all three channels (right panel). Ch1 and the neighbouring Ch2 and Ch3 overlap slightly. The three channel powers are intentionally made different to precompensate the frequency response of the UTC-PD. Reprinted from [J3] © 2013 NPG.

We transmit two kinds of signals over the SISO link. First, in the 'single channel' case (Fig. 4.12a), only the carrier f_1 (Ch1) and the LO f_{LO} are present. We generate QPSK, 8QAM and 16QAM signals with a symbol rate of up to 25 GBd, resulting in bit rates of up to 100 Gbit/s. Second, we discuss the 'multiple channel' case (Fig. 4.12b), with two additional carriers $f_{2,3}$ (Ch2, Ch3) and a fixed spacing Δf_{MLL} = 12.5 GHz (Fig. 4.12b, right). We use sinc-like pulse shaping (see *Section 2.1.5*) with a raised-cosine spectrum and a roll-off factor $\beta = 0.35$ in all channels. Symbol rates are 13 GBd in Ch1, and 8 GBd in Ch2 and Ch3. Channel Ch1 and its neighbors overlap slightly. For our set-up with given comb line spacing and limited system bandwidth, this represents the best tradeoff between digital processing effort in the MFT, linear channel crosstalk and spectral efficiency. At the Tx, we shape the spectrum with the waveshaper to compensate the frequency response of the UTC-PD. With 16QAM signaling in Ch1 and 8QAM signaling in Ch2 and Ch3, the aggregate bit rate is 100 Gbit/s.

4.3.2 Results

The results for the single channel back-to-back (B2B) measurements were already given in in *Section 4.2.2.* Fig. 4.13(a) shows the error vector magnitude (EVM, see *Appendix A.3*) results for the B2B 58 Gbit/s multi-channel transmission with sinc-like pulse shaping and QPSK modulation in all channels. For $P_{Rx} = -28$ dBm, the 13 GBd center channel Ch 1 reaches an EVM value corresponding to a raw bit error ratio (BER) BER $\approx 1 \times 10^{-9}$. The adjacent channels Ch 2 and Ch 3 slightly exceed this value. However, all three channels are clearly below the hard-decision forward error correction (FEC) limit with 7% overhead [171].

Fig. 4.13(b) shows the B2B results for the aggregate 100 Gbit/s multi-channel transmission where Ch 2 and Ch 3 each carry an 8 Gbd 8QAM signal and Ch 1 carries a more demanding 13 GBd 16 QAM signal. The BER of all three channels stays below the 7 % FEC limit, if the receiver input power is larger than about –35 dBm. For the B2B measurements, the setup shown in Fig. A.23 is used.

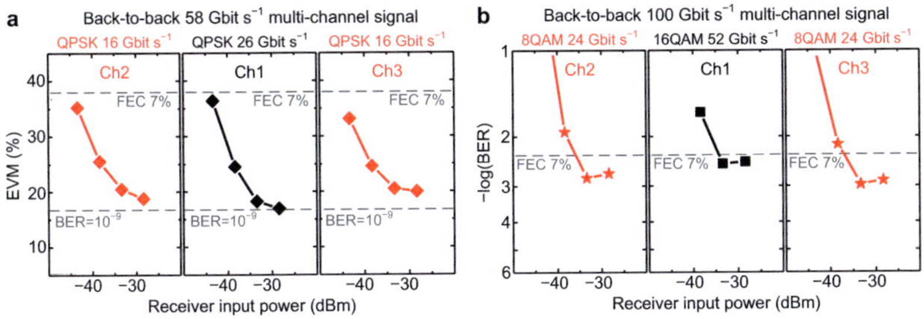

Fig. 4.13: Back-to-back measurement results. (a) EVM for the 58 Gbit/s multi-channel QPSK signal. (b) BER for the 100 Gbit/s multi-channel QAM signal. Reprinted from supplementary information of [J3] © 2013 NPG.

Fig. 4.14(a) shows the EVM of a 50 Gbit/s single-channel QPSK signal for different wireless transmission distances as a function of the relative Rx gain $G = 1/L_{\text{Att}}$, where L_{Att} is the loss added by the Rx waveguide attenuator. The constellation diagram shows clear and distinct symbols. With increasing Rx gain, the EVM decreases (improves) until it starts to show a floor at 16 % (corresponding to a residual BER of $\sim 1 \times 10^{-9}$), which can be seen for $d = 5$, 10 and 20 m. This floor is due to the constant electronic Rx noise contribution. For $d = 40$ m the available receiving power is insufficient to reach this EVM floor. From Fig. 4.14(a) it can also be seen that doubling the transmission distance in the far-field (10…20…40 m) requires four times the receiving power, as would be expected from the free-space path loss.

For 8QAM and 16QAM signaling, Fig. 4.14(b,c), and for $d = 5$ and 10 m, the BER reaches a minimum and begins to increase if the receiving power becomes larger. This behavior is attributed to nonlinearities of the active MMIC Rx that affect multilevel data in particular. This can be seen when comparing to Fig. 4.14(a), where the constant-amplitude QPSK signal does not exhibit an EVM increase for comparable receiving powers.

For the 50 Gbit/s 16QAM signal, we achieve an optimum BER of 3.7×10^{-4} after 40 m. For the 100 Gbit/s 16QAM signal, the optimum BER for distances up to 20 m remains below the FEC limit for 7 % overhead.

Fig. 4.15(a) shows the EVM results for the 58 Gbit/s multiple-channel signal transmission with sinc-like pulse shaping and QPSK modulation in all channels. The 13 GBd centre channel Ch1 performs better than both 8 GBd outer channels Ch2 and Ch3, because the limited receiver bandwidth cuts into the spectra of Ch2 and Ch3.

Fig. 4.15(b) shows the results for the aggregate 100 Gbit/s multiple-channel transmission where Ch2 and Ch3 each carry an 8 GBd 8QAM signal, and Ch1 carries a 13 GBd 16QAM signal. The same receiver bandwidth limitation applies as above, but the signal-to-noise ratio required for transmitting the 16QAM signal is so much larger than for the 8QAM signals that the BER of the center channel Ch1 is worse.

Fig. 4.16 presents the optimum BER from Fig. 4.14(b,c) and Fig. 4.15(b) as a function of the wireless transmission distance. The results for a zero transmission distance are taken from electrical back-to-back measurements. Over distances of $d = 5$, 10 and 20 m, the receiving power is sufficient for an error-corrected transmission, while the BER for $d = 40$ m increases because of the limited power budget.

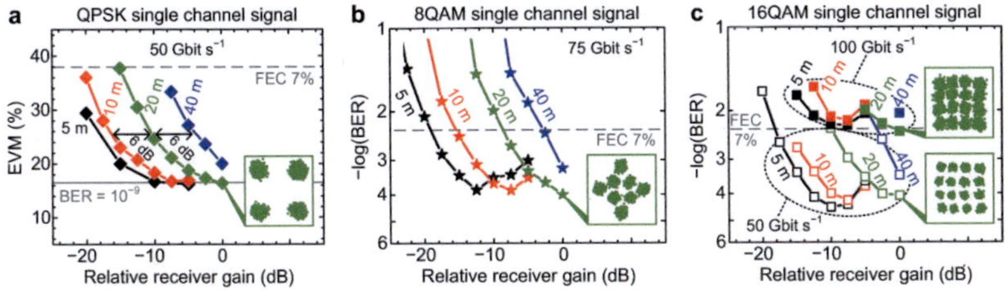

Fig. 4.14: Transmission of single-channel multi-gigabit wireless signals with data rates up to 100 Gbit/s over distances of 5, 10, 20, and 40 m. (a-c), Error vector magnitude (EVM) for received QPSK signals (a) and bit error ratio (BER) for received QAM signals (b,c) as a function of the relative receiver (Rx) gain. The horizontal dashed line indicates the threshold for error-free transmission when using second-generation hard-decision forward error correction (FEC) with 7 % overhead. The insets show the optimum constellation diagrams for 20 m wireless transmission. (a) EVM for the single channel 50 Gbit/s QPSK signal. (b,c) Measured BER for the single channel 75 Gbit/s 8QAM (b) and 50 Gbit/s (□) and 100 Gbit/s (■) 16QAM signals (c). Reprinted from [J3] © 2013 NPG.

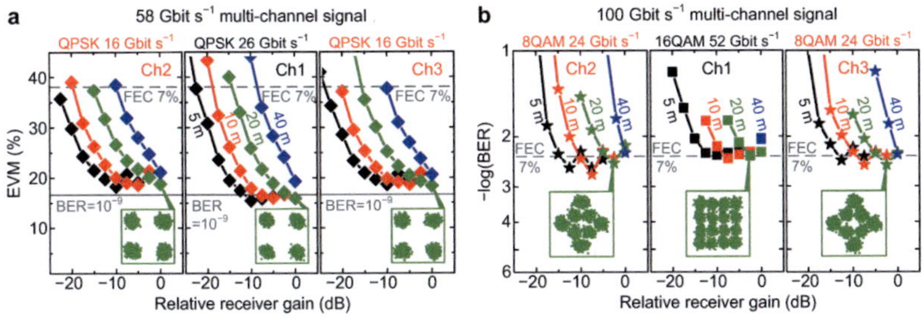

Fig. 4.15: Transmission of multi-channel multi-gigabit wireless signals with data rates up to 100 Gbit/s over distances of 5, 10, 20, and 40 m. Error vector magnitude (EVM) for received QPSK signals (a) and bit error ratio (BER) for received QAM signals (b) as a function of the relative receiver (Rx) gain. The horizontal dashed line indicates the threshold for error-free transmission when using second-generation hard-decision forward error correction (FEC) with 7 % overhead. The insets show the optimum constellation diagrams for 20 m wireless transmission. (a) Measured EVM for 58 Gbit/s multi-channel transmission with QPSK in all three channels. (b) Measured BER for 100 Gbit/s multi-channel transmission with 52 Gbit/s 16QAM in the central channel and 24 Gbit/s 8QAM in the neighboring channels. Reprinted from [J3] © 2013 NPG.

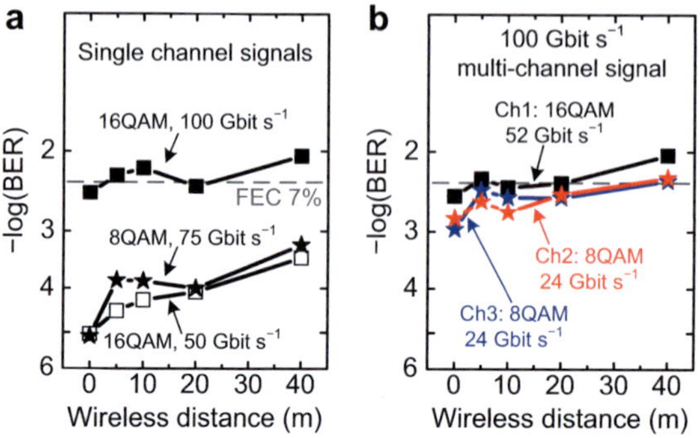

Fig. 4.16: Optimum data transmission as a function of the wireless transmission distance. (a) 8QAM and 16QAM single channel transmission. (b) 100 Gbit/s multi-channel transmission. The results for a zero transmission distance are taken from electrical back-to-back measurements. For distances of 5, 10, and 20 m the receiving power suffices for error-free reception, if an appropriate forward error-correction with 7 % overhead is employed. For a transmission distance of 40 m the BER increases beyond that limit because of the limited power budget. Reprinted from [J3] © 2013 NPG.

4.3.3 Conclusions

In summary, we have demonstrated, for the first time, a SISO wireless link with a maximum data rate of 100 Gbit/s bridging a distance of 20 m using up to three radiofrequency subcarriers near 237.5 GHz. Over 40 m, a maximum data rate of 75 Gbit/s was transmitted with a single radiofrequency carrier at 237.5 GHz and 8QAM signaling. Our results show that combining terahertz photonics and terahertz electronics could pave the way to high-capacity wireless communications using carrier frequencies in the terahertz gap.

By exploiting optical polarization multiplexing and MIMO technology with two photonic wireless transmitters and two MMIC receivers, we could boost the system's performance to a data rate of 200 Gbit/s. A further scaling towards a Tbit/s system could be achieved by taking advantage of optical wavelength division multiplexing, frequency division multiplexing in the 200–300 GHz band and spatial multiplexing with several MIMO links in parallel.

By using MMIC amplifiers [167, 177] at the UTC-PD output and high-gain Cassegrain antennas (>50 dBi) at both the Tx and Rx side, considerably larger point-to-point distances (>1 km) will be possible.

To reach a reliable, compact, pre-commercial system, the integration of the photonic and electronic subcomponents is essential. Recently, the first monolithic IQ electro-optic modulator was demonstrated in gallium arsenide [180], which is also the substrate material of the MMICs in this experiment.

Noting this and anticipating further progress in the field, we foresee that the photonic and electronic subcomponents will be integrated either on a monolithic or hybrid (heterogeneous [181]) platform, leading to compact terahertz transmitters and receivers for wireless communications at Tbit/s rates.

A. Appendix

A.1 Small-Signal Frequency Response of an SOA

The following description of the SOA small-signal modulation response closely follows the derivation given in [S8].

For the SOA small signal-model we assume that there are only small perturbations around an operating point. Then the rate equations can be linearized and solved analytically to study the amplifier's small-signal modulation characteristic.

In a generic formalism of state space representation, a system can be described by two equations, the state equation, Eq. (A.1), and the output equation, Eq. (A.2),

$$\frac{dz}{dt} = \dot{z} = \mathrm{S}\{z(t), y_{\mathrm{in}}(t)\} \tag{A.1}$$

$$y_{\mathrm{out}}(t) = \mathrm{O}\{z(t), y_{\mathrm{in}}(t)\} \tag{A.2}$$

In Eq. (A.1) and Eq. (A.2), z is the state variable, y_{in} is the input variable, and y_{out} is the output variable. The function S describes how the current state z is influenced by previous states and the input variable y_{in}. The function 0 describes how the output variable y_{out} is influenced by the current state z and by the input variable y_{in}.

We assume small perturbations $\Delta z(t)$, $\Delta y_{\mathrm{in}}(t)$, and $\Delta y_{\mathrm{out}}(t)$ around the operating point of the system which is given by z_0, $y_{\mathrm{in},0}$, and $y_{\mathrm{out},0}$. We substitute $z(t)$, $y_{\mathrm{in}}(t)$, and $y_{\mathrm{out}}(t)$ in Eq. (A.1) and Eq. (A.2) by the following three expressions

$$z(t) = z_0 + \Delta z(t) \tag{A.3}$$

$$y_{\mathrm{in}}(t) = y_{\mathrm{in},0} + \Delta y_{\mathrm{in}}(t) \tag{A.4}$$

$$y_{\mathrm{out}}(t) = y_{\mathrm{out},0} + \Delta y_{\mathrm{out}}(t) \tag{A.5}$$

Next we do a Taylor series expansion of Eq. (A.1) and Eq. (A.2) up to the first order, which yields

$$\frac{d\Delta z}{dt} = \left.\frac{\partial \mathrm{S}}{\partial z}\right|_{z=z_0, y_{\mathrm{in}}=y_{\mathrm{in},0}} \cdot \Delta z + \left.\frac{\partial \mathrm{S}}{\partial y_{\mathrm{in}}}\right|_{z=z_0, y_{\mathrm{in}}=y_{\mathrm{in},0}} \cdot \Delta y_{\mathrm{in}} \tag{A.6}$$

$$\Delta y_{\text{out}} = \left.\frac{\partial O}{\partial z}\right|_{z=z_0, y_{\text{in}}=y_{\text{in},0}} \cdot \Delta z + \left.\frac{\partial O}{\partial y_{\text{in}}}\right|_{z=z_0, y_{\text{in}}=y_{\text{in},0}} \cdot \Delta y_{\text{in}} \tag{A.7}$$

The small-signal frequency response of the system is obtained by calculating the Fourier transform of these linearized equations and taking the ratio $\Delta y_{\text{out}}(\omega)/\Delta y_{\text{in}}(\omega)$.

For calculating the SOA small-signal frequency response we start with the rate equation Eq. (2.28) which we write with g from Eq. (2.25) as

$$\frac{\mathrm{d}N}{\mathrm{d}t} = \frac{I}{eV} - \frac{N}{\tau_{\text{c}}} - \frac{g(N)P}{Chf_{\text{s}}} \tag{A.8}$$

Solving Eq. (2.25) for N and inserting the result in Eq. (A.8) yields

$$\frac{\partial g}{\partial t} = \frac{g_{\text{t}} - g}{\tau_c} - \frac{gP}{P_{\text{sat}}\tau_c} \tag{A.9}$$

with P_{sat} from Eq. (2.30) and the new defined constants g_{t} and I_{t}

$$P_{\text{sat}} = \frac{Chf_{\text{s}}}{a\Gamma\tau_{\text{c}}} \tag{A.10}$$

$$g_{\text{t}} = \Gamma a N_{\text{t}}\left(\frac{I}{I_{\text{t}}} - 1\right) \tag{A.11}$$

$$I_{\text{t}} = \frac{qVN_{\text{t}}}{\tau_c} \tag{A.12}$$

Integrating (A.9) over the length L of the active region leads to an expression of the amplifier gain $h(t)$ as a function of the optical input power $P_{\text{in}}(t)$

$$\frac{\partial h}{\partial t} = \frac{g_0 L - h(t)}{\tau_c} - \frac{P_{\text{in}}(t)}{P_{\text{sat}}\tau_c}\left[\exp\{h(t)\} - 1\right] \tag{A.13}$$

From Eq. (2.26) we have

$$\begin{aligned} P_{\text{out}}(t) &= P_{\text{in}}(t)\exp(h) \\ \int_0^L g\,\mathrm{d}z &= h \\ P_{\text{in}}(t) &= P(0,t) \\ P_{\text{out}}(t) &= P(L,t) \end{aligned} \tag{A.14}$$

Note the following useful relations for the integration of (A.9)

$$gP(z,t) = \frac{\mathrm{d}P(z,t)}{\mathrm{d}z}$$

$$\int_0^L gP(z,t)\mathrm{d}z = \int_0^L \frac{\mathrm{d}P(z,t)}{\mathrm{d}z}\mathrm{d}z = P(z,t)\Big|_0^L \tag{A.15}$$

$$P(z,t)\Big|_0^L = P(0,t)\left[\exp(h)-1\right]$$

We apply the state equations (A.6) and (A.7) to Eq. (A.13) and Eq. (A.15) which gives us

$$\frac{\mathrm{d}\Delta h(t)}{\mathrm{d}t} = \underbrace{\left[-\frac{1}{\tau_c} - \frac{P_{\mathrm{in},0}}{P_{\mathrm{sat}}\tau_c}\exp\{h_0\}\right]}_{\mathrm{A}}\Delta h(t) + \underbrace{\left[-\frac{\exp\{h_0\}-1}{P_{\mathrm{sat}}\tau_c}\right]}_{\mathrm{B}}\Delta P_{\mathrm{in}}(t) \tag{A.16}$$

$$\Delta P_{\mathrm{out}}(t) = \underbrace{\left[P_{\mathrm{in},0}\exp\{h_0\}\right]}_{\mathrm{C}}\Delta h(t) + \underbrace{\left[\exp\{h_0\}\right]}_{\mathrm{D}}\Delta P_{\mathrm{in}}(t) \tag{A.17}$$

The state variable z, the input variable y_{in}, and the output variable y_{out} are

$$\begin{array}{ccccc} z & \equiv & h(t) & = & h_0 + \Delta h(t) \\ y_{\mathrm{in}} & \equiv & P_{\mathrm{in}}(t) & = & P_{\mathrm{in},0} + \Delta P_{\mathrm{in}}(t) \\ y_{\mathrm{out}} & \equiv & P_{\mathrm{out}}(t) & = & P_{\mathrm{out},0} + \Delta P_{\mathrm{out}}(t) \end{array} , \tag{A.18}$$

And the output function O and the state function S are

$$\mathrm{O}\{h(t), P_{\mathrm{in}}(t)\} = \frac{g_0 L - h(t)}{\tau_c} - \frac{P_{\mathrm{in}}(t)}{P_{\mathrm{sat}}\tau_c}\left[\exp\{h(t)\}-1\right] \tag{A.19}$$

$$\mathrm{S}\{h(t), P_{\mathrm{in}}(t)\} = P_{\mathrm{in}}(t)\exp\{h(t)\} . \tag{A.20}$$

Taking the Fourier transform of Eq. (A.16) and Eq. (A.17) leads to the two relations

$$\mathrm{j}\omega\Delta h(\omega) = \mathrm{A}\cdot\Delta h(\omega) + \mathrm{B}\cdot\Delta P_{\mathrm{in}}(\omega) \tag{A.21}$$

$$\Delta P_{\mathrm{out}}(\omega) = \mathrm{C}\cdot\Delta h(\omega) + \mathrm{D}\cdot\Delta P_{\mathrm{in}}(\omega). \tag{A.22}$$

where $\omega = 2\pi f$. Solving Eq. (A.21) for $\Delta h(\omega)$, inserting the result into Eq. (A.22), and then solving for the ratio $\Delta P_{\mathrm{out}}(\omega)/\Delta P_{\mathrm{in}}(\omega)$ yields the small-signal frequency responsc $H(\omega)$

$$H(\omega) = \frac{\mathrm{j}\omega\cdot\mathrm{D} + (\mathrm{BC}-\mathrm{AD})}{\mathrm{j}\omega - \mathrm{A}} \tag{A.23}$$

Inserting the expressions for A, B, C, and D from Eq. (A.16) and Eq. (A.17) into Eq. (A.23) gives the final result

$$H(f) = G\frac{\tau_2}{\tau_1}\frac{1+\mathrm{j}2\pi f\tau_1}{1+\mathrm{j}2\pi f\tau_2} \tag{A.24}$$

with the gain G, and two time constants τ_1 and τ_2:

$$G = \exp\{h_0\} \tag{A.25}$$

$$\tau_1 = \tau_\mathrm{c}\left(1+\frac{P_\mathrm{in,0}}{P_\mathrm{sat}}\right)^{-1} \tag{A.26}$$

$$\tau_2 = \tau_\mathrm{c}\left(1+\frac{P_\mathrm{in,0}}{P_\mathrm{sat}}G\right)^{-1} \tag{A.27}$$

The expression (A.24) for the transfer function was also derived in [182]. However, no relations to the physical SOA parameters were given.

A.2 Signal-to-noise Ratio and Noise Figure

This *Section* closely follows the descriptions and derivations given in [183] to calculate the signal-to-noise ratio (SNR) for direct detection and the noise figure (NF) definition for a phase insensitive optical amplifier. The description of the coherent detection and IQ demodulation follows [10].

A.2.1 Direct Reception without Optical Pre-Amplifier

We define the SNR as the ratio γ of the signal power P_S to the noise power P_N

$$\gamma = \frac{P_\text{S}}{P_\text{N}} \tag{A.28}$$

To calculate the SNR, the different noise contributions of the system have to be considered. For direct detection without optical pre-amplifier this is thermal noise and shot noise. Within the bandwidth B and at an absolute temperature T_0, the thermal noise is represented by the noise current $\overline{|i_\text{R}|^2}$ which stands for the thermal noise of the source conductance G_Q and the noise from an electronic pre-amplifier having noise figure F_EA

$$\overline{|i_\text{R}|^2} = 4kT_0 G_\text{Q} F_\text{EA} B \tag{A.29}$$

The letter k denotes Boltzmann's constant. The noise figure F_EA can be expressed as $F_\text{EA} = 1 + F_\text{z}$, where F_z is the excess noise figure which is defined by the ratio T_R/T_0 of the noise temperature T_R and T_0.

The shot noise contribution is given by

$$\overline{|i_\text{shot}|^2} = 2ei_\text{S}B = 2eSP_\text{S}B \tag{A.30}$$

where e is the elementary charge and i_S is the signal current obtained from a photodiode with quantum efficiency η and responsivity S

$$S = \eta \frac{e}{hf} = \eta \frac{e\lambda}{hc} \tag{A.31}$$

In Eq. (A.31), h is Planck's constant, f is the optical frequency, c is the speed of light in vacuum, and λ is the optical wavelength. For an ideal photodetector with maximum quantum efficiency $\eta = 1$, the responsivity is $S = S_1 = e/(hf)$. In practice, the received optical signal power is small and the shot noise is much smaller than the thermal noise. The SNR for direct detection is given by

$$\gamma_{\text{dir}} = \frac{\overline{i_{\text{S}}^2}}{\overline{|i_{\text{R}}|^2} + \overline{|i_{\text{shot}}|^2}} = \frac{\eta P_{\text{S}}}{2hfB} \frac{1}{1 + 4kT_0 G_{\text{Q}} F_{\text{EA}} / (2ei_{\text{S}})} \tag{A.32}$$

The optimum, shot noise limited SNR for direct detection is achieved for large signal photocurrents, when thermal noise becomes negligible compared to shot noise

$$\gamma_{\text{dir}} = \frac{\overline{i_{\text{S}}^2}}{\overline{|i_{\text{shot}}|^2}} = \frac{(SP_{\text{S}})^2}{2eSP_{\text{S}}B} = \frac{\eta P_{\text{S}}}{2hfB} \tag{A.33}$$

In the following, γ_{dir} denotes the shot-noise limited SNR for direct detection without optical pre-amplifier. For a detector quantum efficiency of $\eta = 1$, we write

$$\gamma_1 = \gamma_{\text{dir}}(\eta = 1) = \frac{P_{\text{S}}}{2hfB} \tag{A.34}$$

$\gamma_{\text{dir}}(\eta = 1) = \gamma_1$. As mentioned, in practice the signal power P_{S} is too small to achieve γ_{dir}. With the use of an optical amplifier, one can approach shot-noise limited reception.

A.2.2 Optical Amplifier Noise

Using an optical pre-amplifier with single-pass gain G_{S} increases the signal output power to $G_{\text{S}}P_{\text{S}}$. However, the amplified spontaneous emission (ASE) noise is added by the amplifier. The ASE noise power in one polarization, e.g. linearly polarized in x-direction, within the optical bandwidth B_{O} is given by

$$P_{\text{ASE},x} = (G_{\text{S}} - 1) w_{\text{O}} n_{\text{sp}} B_{\text{O}} \tag{A.35}$$

where $w_{\text{O}} = hf$ is the minimum input noise spectral density (one transverse mode, one polarization), and $n_{\text{sp}} > 1$ is the inversion factor. When receiving the signal P_{S} and the ASE noise $P_{\text{ASE,x}}$ with direct detection within an electrical bandwidth B, the noise comprises the shot noise contribution from P_{S} and $P_{\text{ASE,x}}$

$$\overline{|i_{\text{shot}}|^2} = 2e\left(SG_{\text{S}}P_{\text{S}} + S(G_{\text{S}} - 1) w_{\text{O}} n_{\text{sp}} B_{\text{O}}\right)B, \tag{A.36}$$

the beat noise term that arises from signal-spontaneous noise mixing

$$\overline{|i_{\text{S-sp}}|^2} = 4SG_{\text{S}}P_{\text{S}}S(G_{\text{S}} - 1) w_{\text{O}} n_{\text{sp}} B, \tag{A.37}$$

and the beat term from spontaneous-spontaneous noise mixing

$$\overline{\left|i_{\text{sp-sp}}\right|^2} = 2S^2(G_S-1)^2(w_O n_{sp})^2(B_O - B/2)B. \qquad \text{(A.38)}$$

The SNR at the amplifier's output is then

$$\gamma_{\text{out}} = \frac{i_S^2}{\overline{\left|i_{\text{shot}}\right|^2} + \overline{\left|i_{\text{S-sp}}\right|^2} + \overline{\left|i_{\text{sp-sp}}\right|^2}}$$

$$\gamma_{\text{out}} = \frac{(SG_S P_S)^2}{2e\left(SG_S P_S + S(G_S-1)w_O n_{sp} B_O\right)B + 4SG_S P_S S(G_S-1)w_O n_{sp} B} \qquad \text{(A.39)}$$

$$\overline{\ldots + 2S^2(G_S-1)^2(w_O n_{sp})^2(B_O - B/2)B}$$

A.2.3 Direct Detection with Optical Pre-Amplifier

In practice, $B_O >> B$, $w_O n_{sp} B_O << P_S$, and $G_S >> 1$ are realistic assumptions and the shot noise contribution and the beat term from spontaneous-spontaneous noise mixing can be neglected. The SNR $\gamma_{\text{dir,OA}}$ for direct detection with optical pre-amplifier (neglecting also thermal noise) can be written

$$\gamma_{\text{dir,OA}} = \frac{(SG_S P_S)^2}{4SG_S P_S S(G_S-1)w_O n_{sp} B} \qquad \text{(A.40)}$$

For $w_O n_{sp} = w$ we find

$$\gamma_{\text{dir,OA}} = \frac{1}{2}\frac{P_S}{2w_O B}\frac{1}{(1-1/G_S)} \approx \frac{\gamma_1}{2} \qquad \text{(A.41)}$$

The SNR for direct detection with optical pre-amplifier is about 3 dB smaller compared to the shot noise limited SNR γ_1, which in practice is not achievable for small input powers P_S.

A.2.4 Noise Figure of an Optical Amplifier

The noise characteristics of an amplifier can be described by its noise factor F, or the noise figure $\text{NF} = 10\log_{10}(F)$. The noise factor is defined as the ratio between the $\text{SNR}_{\text{input}}$ at the amplifier's input and the $\text{SNR}_{\text{output}}$ at the amplifier's output

$$F = \frac{\text{SNR}_{\text{input}}}{\text{SNR}_{\text{output}}} \qquad \text{(A.42)}$$

For shot-noise limited reception with an ideal photodetector, the SNR at the amplifier's input is $\mathrm{SNR}_{\mathrm{input}} = \gamma_{\mathrm{dir}}(\eta = 1) = \gamma_1 = P_{\mathrm{S}} / (2hfB) = P_{\mathrm{S}} / (2w_{\mathrm{O}}B)$. With Eq. (A.34) and (A.39), the noise figure can be calculated to

$$\begin{aligned} F &= \frac{\gamma_1}{\gamma_{\mathrm{out}}} \\ F &= \frac{1}{G_{\mathrm{S}}}\left(1 + \frac{G_{\mathrm{S}}-1}{G_{\mathrm{S}}}\frac{w_{\mathrm{O}}n_{\mathrm{sp}}B_{\mathrm{O}}}{P_{\mathrm{S}}}\right) \\ &\quad + 2\frac{G_{\mathrm{S}}-1}{G_{\mathrm{S}}}\frac{w_{\mathrm{O}}n_{\mathrm{sp}}}{w_{\mathrm{O}}} \\ &\quad + \frac{G_{\mathrm{S}}-1}{G_{\mathrm{S}}}\frac{w_{\mathrm{O}}n_{\mathrm{sp}}(B_{\mathrm{O}}-B/2)}{P_{\mathrm{S}}}\frac{G_{\mathrm{S}}-1}{G_{\mathrm{S}}}\frac{w_{\mathrm{O}}n_{\mathrm{sp}}}{w_{\mathrm{O}}} \\ F &= F_{\mathrm{shot}} + F_{\mathrm{S\text{-}sp}} + F_{\mathrm{sp\text{-}sp}} \end{aligned} \tag{A.43}$$

In practice, $B_{\mathrm{O}} >> B$ and $w_{\mathrm{O}}n_{\mathrm{sp}}B_{\mathrm{O}} << P_{\mathrm{S}}$, so that the noise figure becomes independent of the signal input power P_{S}

$$F \approx \frac{1}{G_{\mathrm{S}}} + 2\frac{G_{\mathrm{S}}-1}{G_{\mathrm{S}}}\frac{w_{\mathrm{O}}n_{\mathrm{sp}}}{w_{\mathrm{O}}} \approx F_{\mathrm{shot}} + F_{\mathrm{S\text{-}sp}} \tag{A.44}$$

For $G_{\mathrm{S}} >> 1$ and $n_{\mathrm{sp}} = 1$, the minimum noise figure of a phase-insensitive optical amplifier $F = 2$ (NF = 3dB) is obtained.

A.2.5 Noise Figure of a Cascade of Optical Amplifiers

For calculating the total noise figure of a cascade of optical amplifiers [183], we assume $\eta = 1$ and $n_{\mathrm{sp}} = 1$. Under the assumption $B_{\mathrm{O}} >> B$ and $w_{\mathrm{O}}n_{\mathrm{sp}}B_{\mathrm{O}} << P_{\mathrm{S}}$, only the noise contributions from shot noise and from signal-spontaneous noise mixing are relevant

$$\overline{\left|i_{\mathrm{shot}}\right|^2} + \overline{\left|i_{\mathrm{S\text{-}sp}}\right|^2} = 2eS_1G_{\mathrm{S}}P_{\mathrm{S}}B + 4S_1G_{\mathrm{S}}P_{\mathrm{S}}S_1(G_{\mathrm{S}}-1)w_{\mathrm{O}}B \tag{A.45}$$

We analyze the noise figure of M concatenated optical amplifiers with gain $G_n \geq 1$ ($n = 1...M$) and noise $(G_n - 1)w_{\mathrm{O}}^{(n)}$. Each amplifier is followed by a fiber section with a linear "gain" $0 \leq g_n \leq 1$. The total gain is given by

$$G_{\mathrm{S}} = \prod_{n=1}^{M} G_n g_n \tag{A.46}$$

The SNR $\gamma_{\mathrm{cascade}}$ at the output of the cascade is then given by

$$\gamma_{\text{cascade}} = \frac{(S_1 G_S P_S)^2}{2eS_1 G_S P_S B + 4S_1 G_S P_S S_1 B \times} \cdots$$

$$\cdots \frac{}{\times \left[\begin{array}{l} (G_1 - 1) w_O^{(1)} g_1 \prod_{n=2}^{M} G_n g_n + (G_2 - 1) w_O^{(2)} g_2 \prod_{n=3}^{M} G_n g_n + \ldots \\ + (G_M - 1) w_O^{(M)} g_M \end{array} \right]} \tag{A.47}$$

As can be seen in the denominator of Eq. (A.47), the ASE noise $(G_n - 1)w_O^{(n)}$ of the n-th amplifier is first attenuated by g_n and then experiences the gain of all subsequent amplifier-fiber sections. The noise figure of the cascade is

$$F_M^{(G_n g_n)} = \frac{\gamma_1}{\gamma_{\text{cascade}}}$$

$$F_M^{(G_n g_n)} = \frac{1}{G_S} + 2 \left(\begin{array}{l} \frac{(G_1 - 1)}{G_1} \frac{w_O^{(1)}}{w_O} + \frac{1}{G_1 g_1} \frac{(G_2 - 1)}{G_2} \frac{w_O^{(2)}}{w_O} + \ldots \\ + \frac{1}{\prod_{n=1}^{M-1} G_n g_n} \frac{(G_M - 1)}{G_M} \frac{w_O^{(M)}}{w_O} \end{array} \right)$$

$$F_M^{(G_n g_n)} = \frac{1}{G_S} + \left(F_{\text{S-sp}}^{(1)} + \frac{F_{\text{S-sp}}^{(2)}}{G_1 g_1} + \ldots + \frac{F_{\text{S-sp}}^{(M)}}{\prod_{n=1}^{M-1} G_n g_n} \right) \tag{A.48}$$

Equation (A.48) gives the total noise figure for a sequence of optical amplifiers and fiber sections with the order $G_1 — g_1 — G_2 — g_2 \ldots — G_M — g_M$, i.e., first comes the optical amplifier with gain G_n which is followed by the fiber section g_n. For the case that the fiber section comes before the amplifier, i.e. the sequence $g_1 — G_1 — g_2 — G_2 — \ldots — g_M — G_M$, the total noise figure of the cascade is given by

$$F_M^{(g_n G_n)} = \frac{1}{G_S} + \frac{1}{g_1} \left(F_{\text{S-sp}}^{(1)} + \frac{F_{\text{S-sp}}^{(2)}}{G_1 g_2} + \ldots + \frac{F_{\text{S-sp}}^{(M)}}{\prod_{n=1}^{M-1} G_{n-1} g_n} \right) \tag{A.49}$$

For a chain of identical SOAs with $G_n = G$ compensating the losses $g_n = g$ of the fiber sections ($Gg = 1$), the noise figures from Eq. (A.48) and (A.49) simplify to

$$F_M^{(G_n g_n)} = 1 + MF_{\text{S-sp}} \tag{A.50}$$

$$F_M^{(g_n G_n)} = 1 + MGF_{\text{S-sp}} \tag{A.51}$$

A.2.6 Coherent Reception

For coherent reception, the signal is mixed with a local oscillator (LO) at the photodetector. We write the electric field of the signal $E_S(t)$ as product of a slowly varying amplitude $A_S(t)$ and the optical carrier f_S (phase θ_S) [10]

$$E_S(t) = A_S(t)\exp(j2\pi f_S t + \theta_S) \tag{A.52}$$

For the optical electric field of the LO (frequency f_{LO} and phase θ_{LO} we write

$$E_{LO}(t) = A_{LO}(t)\exp(j2\pi f_{LO} t + \theta_{LO}) \tag{A.53}$$

The amplitudes $A_S(t)$ and $A_{LO}(t)$ are related to the respective signal power P_S and LO power P_{LO} by $P_S = |A_S|^2/2$ and $P_{LO} = |A_{LO}|^2/2$. We assume, that the signal and the LO are co-polarized. The signal and the LO are combined with a 3 dB coupler. The optical electric field incident on each photodiode of a balanced detector is then given by

$$E_1 = \frac{1}{\sqrt{2}}(E_S + E_{LO}) \tag{A.54}$$

$$E_1 = \frac{1}{\sqrt{2}}(E_S - E_{LO}) \tag{A.55}$$

After square law detection the two output photocurrents are given by

$$i_1(t) = SP_1 = \frac{S}{2}\left[P_S + P_{LO} + 2\sqrt{P_S P_{LO}}\cos\left(2\pi f_{IF} t + \theta_S(t) - \theta_{LO}(t)\right)\right] \tag{A.56}$$

$$i_2(t) = SP_2 = \frac{S}{2}\left[P_S + P_{LO} - 2\sqrt{P_S P_{LO}}\cos\left(2\pi f_{IF} t + \theta_S(t) - \theta_{LO}(t)\right)\right] \tag{A.57}$$

where $f_{IF} = f_S - f_{LO}$ is the intermediate frequency (IF). The total output photocurrent of the balanced detector is obtained by taking the difference $i_1 - i_2$ (balanced detection)

$$i(t) = i_1(t) - i_2(t) = 2S\sqrt{P_S P_{LO}}\cos\left(2\pi f_{IF} t + \theta_S(t) - \theta_{LO}(t)\right) \tag{A.58}$$

As can be seen from Eq. (A.58), the DC terms SP_S and SP_{LO} from Eq. (A.56) and (A.57) are eliminated by balanced detection.

A.2.7 Coherent Reception with a 90° Optical Hybrid

When using a 90° optical hybrid, see Fig. 2.9, for coherent reception, we can write for the four electric optical output fields [10]

$$E_1(t)=\frac{1}{2}(E_S+E_{LO}) \tag{A.59}$$

$$E_2(t)=\frac{1}{2}(E_S-E_{LO}) \tag{A.60}$$

$$E_3(t)=\frac{1}{2}(E_S+E_{LO}) \tag{A.61}$$

$$E_4(t)=\frac{1}{2}(E_S-E_{LO}) \tag{A.62}$$

Note the factor ½ compared to the factor $1/\sqrt{2}$ in Eq. (A.54) and (A.55) due to the 3 dB couplers in the 90° optical hybrid. An analogy to the coherent detection without 90° optical hybrid, we can calculate the resulting inphase current $i_I(t)$ and the quadrature current $i_Q(t)$ after balanced detection

$$i_I(t)=i_1(t)-i_2(t)=S\sqrt{P_SP_{LO}}\cos\left(2\pi f_{IF}t+\theta_S(t)-\theta_{LO}(t)\right) \tag{A.63}$$

$$i_Q(t)=i_3(t)-i_4(t)=S\sqrt{P_SP_{LO}}\sin\left(2\pi f_{IF}t+\theta_S(t)-\theta_{LO}(t)\right) \tag{A.64}$$

The complex signal $i_C(t)$ is then given by

$$i_C(t)=i_I(t)+\mathrm{j}i_Q(t)=S\sqrt{P_SP_{LO}}\exp\left(\mathrm{j}\left(2\pi f_{IF}t+\theta_S(t)-\theta_{LO}(t)\right)\right) \tag{A.65}$$

For homodyne phase locked reception the signal becomes

$$i_C(t)=i_I(t)+\mathrm{j}i_Q(t)=S\sqrt{P_SP_{LO}}\exp\left(\mathrm{j}\theta_S(t)\right) \tag{A.66}$$

and the complex signals can directly be recovered.

A.2.8 Optical Heterodyning

Optical heterodyning is an elegant photonic technique to generate wireless signals. Here we calculate the SNR for the optical heterodyne reception. In order to make the results directly comparable to the SNR results obtained for direct detection, we assume that there is no 3 dB coupler in front of the photodetector and that the photodetector is a single photodiode and no balanced detector. The electric optical field of the signal and LO on the photodiode are then given by Eq. (2.19) and (2.20). The resulting photocurrent after photodetection is

$$i(t)=S\left\lfloor P_S+P_{LO}+2\sqrt{P_SP_{LO}}\cos\left(2\pi f_{IF}t+\theta_S(t)-\theta_{LO}(t)\right)\right\rfloor \tag{A.67}$$

We are interested in the generated signal at the IF f_{IF}. The average signal current $\overline{i_S^2}$ is given by

$$\overline{i_{\mathrm{S}}^2} = \overline{\left[2S\sqrt{P_{\mathrm{S}}P_{\mathrm{LO}}}\cos\left(2\pi f_{\mathrm{IF}}t + \theta_{\mathrm{S}}(t) - \theta_{\mathrm{LO}}(t)\right)\right]^2} = 2S^2 P_{\mathrm{S}} P_{\mathrm{LO}} \qquad \text{(A.68)}$$

We assume that $P_{\mathrm{LO}} >> P_{\mathrm{S}}$, thus the shot noise contribution will be dominated by the current $i_{\mathrm{LO}} = SP_{\mathrm{LO}}$. At the IF, the signal occupies the bandwidth $2B$. For optical heterodyning we write for the SNR at the IF

$$\gamma_{\mathrm{Het,IF}} = \frac{\overline{i_{\mathrm{S}}^2}}{\overline{|i_{\mathrm{R}}|^2} + \overline{|i_{\mathrm{shot}}|^2}} = \frac{2S^2 P_{\mathrm{S}} P_{\mathrm{LO}}}{4kT_0 G_{\mathrm{Q}} F_{\mathrm{EA}} 2B + 2eP_{\mathrm{LO}} S 2B}$$
$$\gamma_{\mathrm{Het,IF}} = \gamma_{\mathrm{dir}} \frac{1}{1 + 4kT_0 G_{\mathrm{Q}} F_{\mathrm{EA}} / (2ei_{\mathrm{LO}})} \qquad \text{(A.69)}$$

For $i_{\mathrm{LO}} = SP_{\mathrm{LO}} >> 2kT_0 G_{\mathrm{Q}} F_{\mathrm{EA}} / e$, the maximum *baseband* shot-noise limited SNR $\gamma_{\mathrm{Het,BB}}$ is given by

$$\gamma_{\mathrm{Het,BB}} = 2\gamma_{\mathrm{Het,IF}} = 2\gamma_{\mathrm{dir}} \qquad \text{(A.70)}$$

A.3 Q-Factor, Bit Error Ratio, Error Vector Magnitude

The content of Section A.3.1 has been published in the conference publication [C15]:

W. Freude, R. Schmogrow, B. Nebendahl, M. Winter, A. Josten, D. Hillerkuss, **S. Koenig**, J. Meyer, M. Dreschmann, M. Huebner, C. Koos, J. Becker, and J. Leuthold, "Quality metrics for optical signals: Eye diagram, Q-factor, OSNR, EVM and BER," *14th International Conference on Transparent Optical Networks (ICTON)*, Coventry, United Kingdom, July 2012, paper Mo.B1.5 (invited), doi: 10.1109/ICTON.2012.6254380.

The ultimate quality measure in optical communication links is the bit error ratio (BER, bit error probability). However, a direct experimental BER determination can be done only as long as the link is out of service (if not a hard-decision forward error correction (FEC) is able to report errors), because a known data sequence has to be transmitted, and it consumes a significant amount of time if the BER is small. If the estimated $\mathrm{BER} = 10^{-r}$ (which itself is a random number) should lie with probability $\alpha = 99\%$ in a confidence interval $\varepsilon = (\mathrm{BER}_{\max} - \mathrm{BER}) / \mathrm{BER} = 100\ \%$ about 13×10^{r} bit errors have to be recorded [184]. For a data rate R, a single BER measurement takes the time $\tau_{\mathrm{BER}} = 13 \times 10^{r}\ \mathrm{bit}/R$. For $r = 4$ ($r = 12$) and $R = 10$ Gbit/s we find $\tau_{\mathrm{BER}} = 13\ \mu\mathrm{s}$ ($\tau_{\mathrm{BER}} = 1300$ s). For simulations and off-line signal processing the effective data rate is smaller by many orders of magnitude, so counting errors becomes impractical. It is therefore tempting to estimate the BER based on a few measured moments of the probability density function (pdf), which describes the noise at the decision circuit of the receiver. In the following, we discuss various moment-based metrics and compare simulations with experiments.

A.3.1 Eye Diagram and Q-Factor

A fairly general receiver schematic for on-off-keying (OOK) signals is depicted in Fig. A.1(a) (top). It consists of an optical pre-amplifier (OA) and a filter with a bandwidth B_{O} having an output electric field strength E. A photodetector transforms the time average $<|E|^2>$ over a few optical periods into a proportional photocurrent, which is converted into a voltage by an electronic amplifier (EA). After a filter with bandwidth B_{E} the output voltage u has to be associated with a one or a zero.

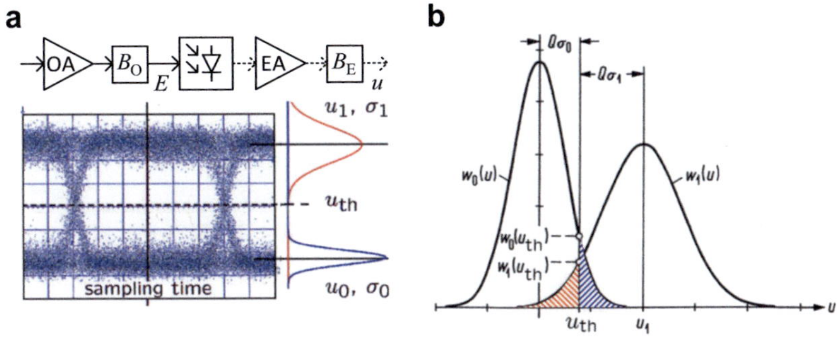

Fig. A.1: (a) Direct OOK receiver and eye diagram. **(top)** Optical pre-amplifier (OA), filter (bandwidth B_{O}) and electrical field E. Photodiode with electrical amplifier (EA), filter (bandwidth B_{E}) and output voltage u. **(bottom)** Eye diagram of voltage u with schematic noise histograms (pdf) at mean one level u_1 (standard deviation σ_1, red) and mean zero level u_0 (σ_0, blue), [C15]. (b) Probability density functions $w_{0,1}(u)$ of output noise voltage at mean one and zero levels $u_{1,2}$ (variances $\sigma_{1,2}$) with decider threshold voltage u_{th}. The hatched blue area ($u > u_{\mathrm{th}}$) represents the probability $p(1\mathrm{r}|0\mathrm{t})$ to actually receive a one if a zero was transmitted, while the hatched red area ($u < u_{\mathrm{th}}$) is the probability $p(0\mathrm{r}|1\mathrm{t})$ to receive a zero if a one was transmitted. Reprinted from [C15] © 2012 IEEE.

First, we look at a receiver without optical pre-amplifier. For small received signal powers powers $<|E|^2>$ the noise is mostly due to the EA. This situation is depicted in form of an eye diagram of the voltage u as a function of time t. The eye shows very little intersymbol interference (ISI). Detailed methods to analyze the eye were reported recently [185]. We operate at the optimum sampling time, Fig. A.1(a) (bottom). Significant noise will be recorded. According to the measured histograms the fluctuating voltage is distributed symmetrically around the average values u_1 for a one and u_0 for a zero. The probability density functions (pdf) of these noise voltages are depicted in Fig. A.1(b) and are denoted with $w_1(u)$ and $w_0(u)$, respectively. The probability of transmitting a one or a zero is $p(1\mathrm{t})$ or $p(0\mathrm{t})$, and the BER corresponds to the sum of the appropriately weighed hatched areas in Fig. A.1(b). The optimum decision threshold $u_{\mathrm{th,op}}$ for a minimum $\mathrm{BER_{op}}$ is found by calculating $\mathrm{d\,BER}/\mathrm{d}u_{\mathrm{th}} = 0$,

$$\mathrm{BER} = p(1\mathrm{t})\int_{-\infty}^{u_{\mathrm{th}}} w_1(u)\,\mathrm{d}u + p(0\mathrm{t})\int_{u_{\mathrm{th}}}^{+\infty} w_0(u)\,\mathrm{d}u, \quad p(1\mathrm{t})w_1(u_{\mathrm{th,op}}) = p(0\mathrm{t})w_0(u_{\mathrm{th,op}}) \quad \text{(A.71)}$$

If ones and zeros are transmitted with equal probability, $p(1\mathrm{t}) = p(0\mathrm{t})$, and the pdf are unimodal, the intersection of $w_1(u)$ and $w_0(u)$ marks the optimum threshold. For an estimate of the BER, both pdf have to be known, and if the BER is small, then the pdf must be accurately known especially in the wings. If the EA noise dominates, the statistics can be well approximated by two Gaussian pdf with mean $u_{1,0}$ and standard deviation $\sigma_{1,0}$, namely $w_{1,0}(u) = 1/(\sqrt{2\pi}\,\sigma_{1,0})\exp[(u-u_{1,0})^2/(2\sigma_{1,0}^2)]$. This approximation is well justified by the central limit theorem [186], which states that the pdf of a sum of independent random variables (represented by the output of filter B_{E}) tends towards a Gaussian. The complementary error function $\mathrm{erfc}(z) = (2/\sqrt{\pi})\int_z^{\infty}\exp(-t^2)\,\mathrm{d}t$ allows to formulate a simple analytic expression for the optimum $\mathrm{BER_{op}}$, if the ratio of the occurrence probabilities for ones and zeros equals the ratio of the corresponding standard deviations,

$$\begin{gathered}\mathrm{BER_{op}} = \frac{1}{2}\mathrm{erfc}\left(\frac{Q}{\sqrt{2}}\right), \quad Q = \frac{u_1 - u_0}{\sigma_1 + \sigma_0} \approx \sqrt{\gamma}, \\ u_{\mathrm{th}} = \frac{u_1\sigma_0 + u_0\sigma_1}{\sigma_0 + \sigma_1}, \quad \frac{w_1(u_{\mathrm{th,op}})}{\sigma_0} = \frac{w_0(u_{\mathrm{th,op}})}{\sigma_1} \quad \text{for} \quad \frac{p(1\mathrm{t})}{p(0\mathrm{t})} = \frac{\sigma_1}{\sigma_0}\end{gathered} \quad \text{(A.72)}$$

With a sampling oscilloscope we record the random voltage u at the given (optimum) sampling time. From these data the four moments $u_{1,0}$ and $\sigma_{1,0}$ are computed, so that $u_{\mathrm{th,op}}$ and $\mathrm{BER_{op}}$ can be estimated from Eq. (A.72). It is worth noting that $u_{1,0}$ and $\sigma_{1,0}$, being calculated from a statistical sample only, are themselves random variables, and this property is inherited by the

BER estimate. All measured quantities involved are either first moments $u_{1,0}$ or central second moments $\sigma^2_{1,0}$ of the random voltage sample u in Fig. A.1(a).

The so-called Q-factor in Eq. (A.72), a quality metric that can be defined by moments up to the order two (even if the actual pdf are not known), is related to the electrical signal-to-noise power ratio $\gamma = P_{S,a}/P_N$ under the following assumptions: The received pulses have no ISI, ones and zeros are uniformly distributed, a zero has the mean value $u_0 = 0$, the relation $1 < \sigma_1^2/\sigma_0^2 \le 2$ holds, the average noise power $P_N = \left(\sigma_1^2 + \sigma_0^2\right)/2$ is limited to $\sigma_1^2 < P_N < 1.5\sigma_1^2$. The signal power of *one* typical voltage impulse shape with maximum $h(0) = 1$ (e. g., a raised cosine) is $P_S^{(1)} = \left\langle u_1^2 h^2(t)\right\rangle \approx u_1^2/2$, which leads to an *average* signal power $P_{S,a} = P_S^{(1)}/2 \approx u_1^2/4$. We then find $\gamma \approx (\frac{1}{3}\ldots\frac{1}{4})u_1^2/\sigma_1^2$ and $Q^2 \approx (\frac{1}{2.91}\ldots\frac{1}{4})u_1^2/\sigma_1^2$ resulting in $\gamma \approx (0.97\ldots1)Q^2 \approx Q^2$ [187]. Therefore a (not very accurate) BER estimate can be derived from a measured electrical signal-to-noise power ratio $\gamma \approx Q^2$.

A.3.2 Error Vector Magnitude

The content of Section A.3.2 has been published in [J6]. Minor changes have been made to adjust the notations of variables.

R. Schmogrow, B. Nebendahl, M. Winter, A. Josten, D. Hillerkuss, **S. Koenig**, J. Meyer, M. Dreschmann, M. Huebner, C. Koos, J. Becker, W. Freude, and J. Leuthold, "Error Vector Magnitude as a Performance Measure for Advanced Modulation Formats," *IEEE Photon. Technol. Lett.*, vol. 24, no. 1, pp. 61-63, 2012, doi: 10.1109/LPT.2011.2172405. Correction: ibid., vol. 24, no. 23, p. 2198, 2012, doi: 10.1109/LPT.2012.2219471.

Coherent optical transmission systems and advanced modulation formats such as M-ary quadrature amplitude modulation (QAM) are establishing quickly [188]. To encode these formats a variety of new optical modulator concepts have been introduced [189]. Among them are modulators dedicated to a particular modulation format [103] as well as novel software-defined optical transmitters that allow encoding of many modulation formats at the push of a button [83, 88]. In light of the capabilities to encode such advanced modulation formats there is a need to reliably judge the quality of the encoded signals. In laboratory experiments so far most receivers employ offline digital signal processing (DSP) at much reduced clock rates. This offline processing makes it very time consuming to reliably compute the bit error ratio (BER), especially if the signal quality is high. As a consequence, a faster – yet reliable – performance measure is needed, in particular when investigating wavelength division multiplexing (WDM) [190] or multi-carrier systems [J7].

Traditionally, the Q-factor metric is well established for on-off keying (OOK) optical systems. To estimate BER from Q, marks and spaces in the detected photocurrent are assumed to be superimposed with additive white Gaussian noise (AWGN), the probability density of which is fully described by its mean and variance. A large Q leads to a small BER.

Unfortunately, the method cannot be simply transferred to QAM signals, where the optical carrier is modulated with multi-level signals both in amplitude and phase. Instead, the error vector magnitude (EVM) is employed. It describes the effective distance of the received complex symbol from its ideal position in the constellation diagram. If the received optical field is perturbed by AWGN only, the EVM can be related to BER and to the optical signal-to-noise ratio (OSNR) [191, 192]. A small EVM leads then to a small BER. The EVM metric is standard in wireless and wireline communications. However, its connection to BER and OSNR is not well established in optical communications. Especially one has to discriminate between data-aided reception, where for measurement purposes the actually sent data are known, as opposed to nondata-aided reception, where the received data are unknown. The first case is standard for BER measurements, while the second case is more common for real-world receivers (disregarding, e.g., training sequences). For strongly noisy signals, nondata-aided reception tends to underestimate the EVM, because a received symbol could be nearer to a "wrong" constellation point than to its "right" position.

In this paper we confirm experimentally and by simulations that the BER can be estimated from EVM data by an analytic relation [191]. Strictly speaking, this BER estimate is valid for data-aided reception only, but we found that the method can be also applied for nondata-aided reception if $\text{BER} < 10^{-2}$ holds. Further, the EVM can be estimated [192] if the OSNR has been measured. Both estimates are valid for systems limited by optical AWGN. To support our findings we compare measured OSNR, EVM and BER for symbol rates of 20 GBd and 25 GBd with calculated BER and EVM estimates for the modulation formats binary phase shift keying (BPSK), quadrature PSK (QPSK), 8PSK, 16QAM, 32QAM, and 64QAM.

EVM Definition

Advanced modulation formats such as M-ary QAM encode a data signal in amplitude and phase of the optical electric field. The resulting complex amplitude of this field is de described by points in a complex constellation plane. Fig. A.2(a) depicts the ideal constellation points for a 16QAM signal. The actually received signal vector E_r deviates by an error vector E_{err} from the ideal transmitted vector E_{t}. Fig. A.2(b) shows a simulated noisy constellation.

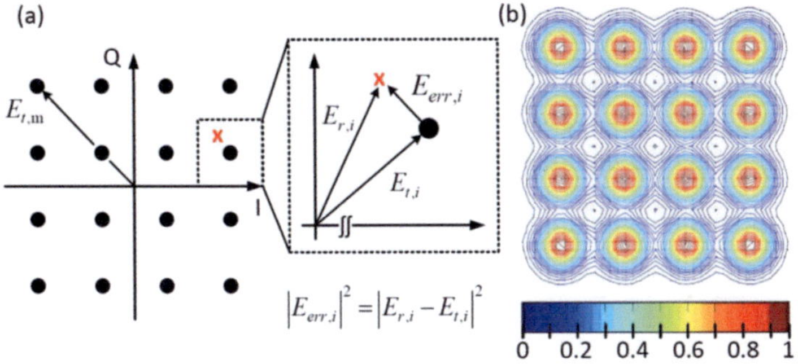

Fig. A.2: Constellation diagram and error vector for a 16QAM signal. (**a**) Ideal constellation diagram with an actually transmitted value x. The blow-up illustrates the definition of the ith error vector $E_{\mathrm{err},i}$ in relation to the actually received signal vector $E_{r,i}$ and the vector $E_{t,i}$ of the transmitted signal. (**b**) Simulated constellation diagram with white Gaussian noise, EVM = 15 %. Color coding for reception probability of transmitted symbols. Reprinted from [J6] © 2012 IEEE.

The EVM is defined by a root mean square of E_{err} for a number of I randomly transmitted data [191] and embraces all (linear and nonlinear) impairments:

$$\mathrm{EVM_m} = \frac{\sigma_{\mathrm{err}}}{|E_{t,\mathrm{m}}|},\ \sigma_{\mathrm{err}}^2 = \frac{1}{I}\sum_{i=1}^{I}|E_{\mathrm{err},i}|^2,\ E_{\mathrm{err},i} = E_{r,i} - E_{t,i} \tag{A.73}$$

The power of the longest ideal constellation vector with magnitude $|E_{t,\mathrm{m}}|$ serves for normalization. Other authors use the average power $|E_{t,\mathrm{a}}|^2$ of all M symbol vectors within a constellation leading to $\mathrm{EVM_a}$. The two EVM normalizations are related by a modulation format-dependent factor k ,

$$\mathrm{EVM_a} = k\,\mathrm{EVM_m},\ k^2 = \frac{|E_{t,\mathrm{m}}|^2}{|E_{t,\mathrm{a}}|^2},\ |E_{t,\mathrm{a}}|^2 = \frac{1}{M}\sum_{i=1}^{M}|E_{t,i}|^2. \tag{A.74}$$

Table A.1 specifies the k-values relating the two definitions for the modulation formats discussed here.

	B/Q/8PSK	16QAM	32QAM	64QAM
k^2	1	9/5	17/10	7/3

Table A.1: Modulation format dependent factor k^2 from Eq. (A.74).

Relations between OSNR, EVM, and BER

The EVM_m from Eq. (A.73) can be estimated from OSNR (measured for instance with an optical spectrum analyzer, OSA) [191] according to Eq. (A.75). The basic assumptions are that system errors are mainly due to optical AWGN (neglecting nonlinear effects and electronic noise), that reception is nondata-aided, and that quadratic M-QAM signal constellations are regarded. With Eq. (A.74) we find from [192]:

$$\mathrm{EVM}_\mathrm{m} \approx \frac{1}{k}\left[\begin{array}{l}\dfrac{1}{\mathrm{OSNR}} - \sqrt{\dfrac{96}{\pi(M-1)\mathrm{OSNR}}}\displaystyle\sum_{i=1}^{\sqrt{M}-1}\gamma_i \mathrm{e}^{-\frac{3\beta_i^2\mathrm{OSNR}}{2(M-1)}} \\ \quad + \dfrac{12}{M-1}\displaystyle\sum_{i=1}^{\sqrt{M}-1}\gamma_i\beta_i\,\mathrm{erfc}\left(\sqrt{\dfrac{3\beta_i^2\mathrm{OSNR}}{2(M-1)}}\right)\end{array}\right]^{\frac{1}{2}}, \tag{A.75}$$

$$\gamma_i = 1 - i/\sqrt{M}, \quad \beta_i = 2i-1.$$

The first term in Eq. (A.75), i.e. $\mathrm{EVM}_\mathrm{m} \approx 1/\left(k\sqrt{\mathrm{OSNR}}\right)$, rewrites Eq. (A.73) and (A.74) for the case of data-aided reception, and if optical AWGN is the dominant source of E_err. The remaining terms account for nondata-aided reception and disappear for large OSNR. For large numbers of constellation points M only the first few terms in the summation need to be considered. To estimate a BER from EVM_m we define L as the number of signal levels identical within each dimension of the (quadratic) constellation ($L^2 = M$), and $\log_2 M$ as the number of bits encoded into each QAM symbol. The BER is then approximated by [191]

$$\mathrm{BER} \approx \frac{\left(1-L^{-1}\right)}{\log_2 L}\mathrm{erfc}\left[\sqrt{\frac{3\log_2 L}{\left(L^2-1\right)}\frac{1}{\left(k\,\mathrm{EVM}_\mathrm{m}\right)^2\log_2 M}}\right] \tag{A.76}$$

For Eq. (A.76), the same limitations as with Eq. (A.75) apply, but in this case data-aided reception is assumed. If the EVM_m is not derived by evaluating Eq. (A.75) but measured directly, the influence of electronic noise is also included.

The EVM and the Q^2-factor are related. In direct detection OOK systems assuming *electrical* AWGN with a standard deviation $\sigma_1 \propto \sigma_Q^2$ for the photocurrent $i_1 \propto |E_{t,\mathrm{m}}|^2$ of a mark, the Q-factor in the shot-noise limited case, $\sigma_0 \approx 0$, is defined in analogy to Eq. (A.73) by

$$Q = \frac{i_1}{\sigma_1 + \sigma_0} \approx \frac{|E_{t,\mathrm{m}}|^2}{\sigma_Q^2}, \quad \sigma_Q^2 = \frac{1}{I}\sum_{i=1}^{I}\left[|E_{r,i}|^2 - |E_{t,i}|^2\right] \tag{A.77}$$

The Q^2-factor represents an *electrical* signal-to-noise power ratio and provides for OOK signals a good estimate of the $\mathrm{BER} \approx (1/2)\operatorname{erfc}\left(Q/\sqrt{2}\right)$. Conversely, the EVM is based on electrical fields and thus assesses the BER for a variety of formats accounting for both *optical* and *electrical* AWGN.

In the following, we compare the theoretical predictions Eq. (A.75) and (A.76) with numerical simulations and measurements.

Experimental Setup

We measure OSNR, EVM and BER in a software-defined real-time multi format transmitter setup [83], Fig. A.3. We sequentially generate the six complex modulation formats B/Q/8PSK and 16/32/64QAM at symbol rates of 20 GBd and 25 GBd on an external cavity laser (ECL) at 1550 nm. The modulated carrier is kept at a fixed average power and combined with a variable noise source (amplified spontaneous emission (ASE) source with attenuator) to vary the OSNR. An optical spectrum analyzer (OSA) determines the amplified signal's OSNR. An Agilent modulation analyzer (OMA) decodes the modulation and measures EVM and BER. The insets of Fig. A.3 display the spectrum of the ASE source as well as constellations at 25 GBd of three selected formats and at best available OSNR.

Experimental Results

In Fig. A.4(a) we display the measured EVM_m for various measured OSNR values (closed symbols for 20 GBd, open symbols for 25 GBd). The solid lines represent Eq. (A.75). For OSNR < 20 dB the theoretical prediction coincides with the measurement. Constellations of *x*QAM can be recovered for OSNR > 12 dB only. For OSNR > 20 dB the electronic noise dominates so that Eq. (A.75) does not hold any more. If the electronic noise contribution would be less, as is the case for systems with lower symbol rate and consequently smaller bandwidth, the error floor would be seen at higher OSNR only.

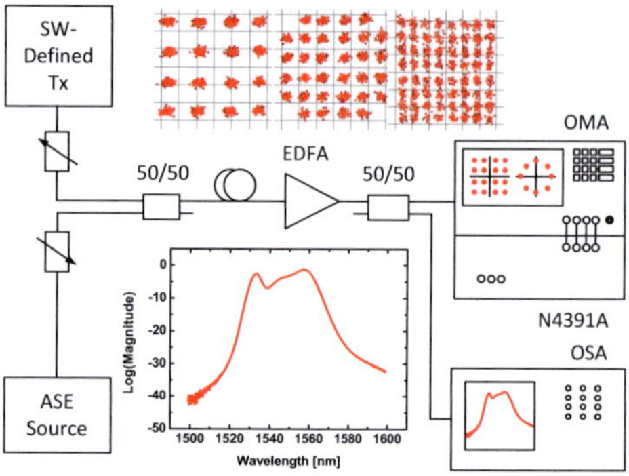

Fig. A.3: Experimental setup for BER and EVM measurements. The optical signal of a software-defined transmitter [83] generates a choice of six different modulation formats (three of which are shown) for optical signal-to-noise ratios adjusted by injecting a variable amount of amplified spontaneous emission (ASE). After amplification with an erbium-doped fiber amplifier (EDFA), the OSNR is measured by an optical spectrum analyzer (OSA). The modulation is decoded by an Agilent optical modulation analyzer (OMA). Reprinted from [J6] © 2012 IEEE.

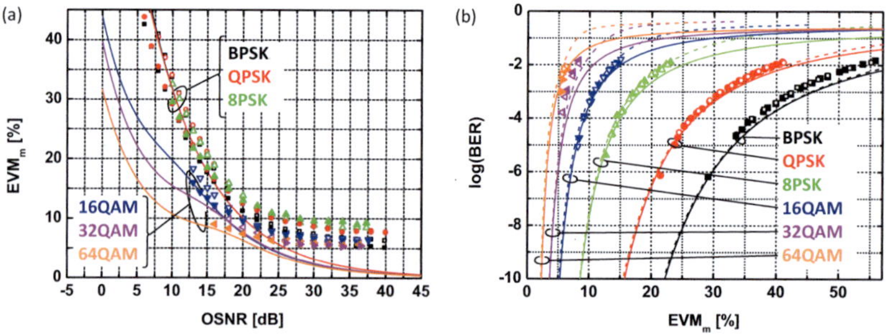

Fig. A.4: Interdependencies of OSNR, EVM_m and BER. Filled symbols represent measurements for a symbol rate of 20 GBd, open symbols for 25 GBd. (a) Measured (symbols) and calculated [191] EVM_m (solid lines) as a function of OSNR. For high OSNR levels the measured plots have an error floor due to the electronic noise of the transmitter and receiver. The different error floors for Q/8PSK and *x*QAM stem from different factors *k*. The error floor for BPSK is lower because of transmitter specific properties. (b) Measured (symbols), simulated (dashed lines) and calculated BER (solid lines) as a function of EVM_m. Reprinted from [J6] © 2012 IEEE.

Fig. A.4(b) shows the measured BER as a function of the measured EVM (closed symbols: 20 GBd, open symbols: 25 GBd). The solid lines represent Eq. (A.76), the dashed lines result from simulations. While measurement and simulation are based on nondata-aided reception, Eq. (A.76) assumes data-aided detection. Still, measurement, analytical estimate and simulations coincide for a large range up to a BER of 10^{-2}.

Some more information can be extracted from Fig. A.4(a) and (b). While the 32QAM constellation is not strictly quadratic, it is nearly so, and hence the estimation quality is comparable to the one for the quadratic formats. The plots also show that the EVM depends on the format, as higher-order formats are more sensitive to noise than others, as predicted by Eq. (A.75) and Eq. (A.76).

For determining the BER we use a $2^{15}-1$ pseudo random binary sequence. The number of compared bits and the number of recorded errors were chosen according to the statistical reasoning described in [193].

A.4 Millimeter-Wave Monolithic Integrated Circuit Mixers

In the following *Subsections*, the electronic mixers that are used for electronic up- and down-conversion in the experiments of *Chapter 4* are described. The mixers were fabricated at the Fraunhofer Institute of Applied Solid-State Physics Freiburg (IAF).

A.4.1 Electronic Mixers for Real-Valued Signals

Active mm-wave monolithic integrated circuits (MMIC) based on metamorphic high electron mobility transistor (mHEMT) technology with a gate length of 50 nm were used to implement the electronic Tx and Rx mixers. With this technology, heterostructures with a very high electron mobility of 11800 $cm^2/(Vs)$ are achieved in a InGaAs channel which is grown on top of a metamorphic buffer with linear InAlGaAs grading on a GaAs wafer. The obtained transit frequency and maximum oscillation frequency are f_T = 380 GHz and f_{max} > 600 GHz [167].

Fig. A.5 (a,b) show the functional blocks of the single-chip Tx and Rx MMICs that are used in Section 4.1 for electronic up-conversion to the carrier frequency of 220 GHz and down-conversion to baseband, respectively. In both the Tx and the Rx MMIC, the frequency conversion is realized by a resistive mixer, whose LO signal is provided via an integrated frequency doubler stage. The Tx mixer's signal input port and the Rx mixer's output port for the down-converted signal are denoted as "IF port". In the Tx, a four-stage cascode low-noise amplifier (LNA) amplifier with 20 dB small-signal gain and an estimated noise figure of 7.9 dB is used as post-amplifier to amplify the mixer's RF output signal. The RF output power of the Tx MMIC is about 0 dBm. The same type of four-stage LNA amplifies the received mm-wave signal at the input of the Rx MMIC. The conversion gain of the Rx module is > −4.8 dB. The chip dimensions of both the Tx and Rx MMIC are 0.75×2 mm^2. Further information on the Tx and Rx MMIC technology and the module characteristics can be found in [41, 168].

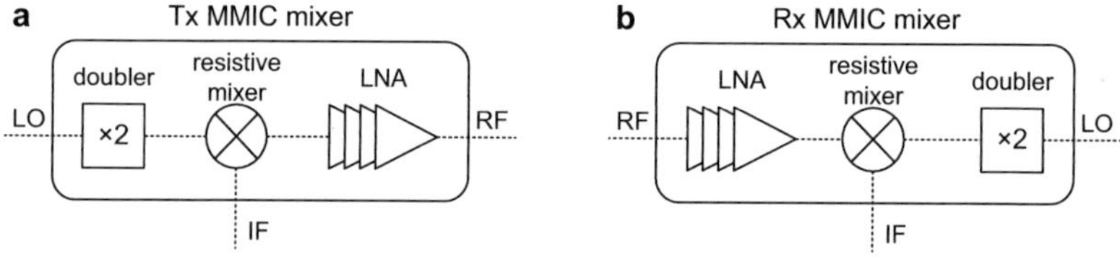

Fig. A.5: Block diagram of the Tx MMIC (a) and the Rx MMIC (b) used for electronic up- and down-conversion of real-valued data signals. The MMICs comprise a identical components: a local oscillator (LO) frequency doubler, a resistive mixer, and a low-noise amplifier (LNA) stage.

A.4.2 Electronic Mixers for Complex-Valued IQ-Signals

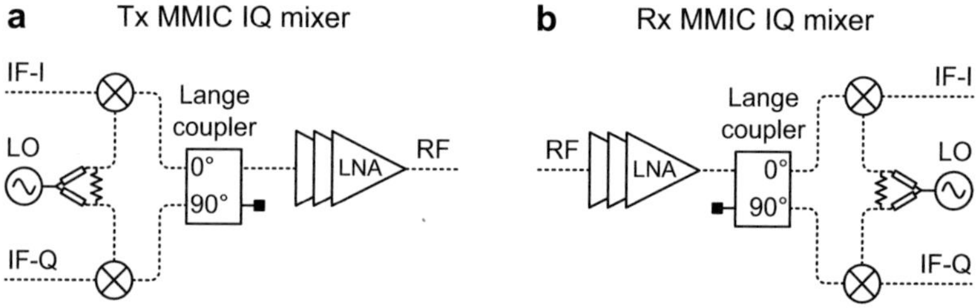

Fig. A.6: Schematic block diagram of the Tx MMIC IQ mixer (a) and the Rx MMIC IQ mixer (b). Note that the 90° phase shift for quadrature modulation and demodulation is provided by the Lange coupler and not by a 90° phase shifter element in the LO path.

The active MMIC Tx and Rx IQ mixers employed in the experiments of this thesis are fabricated in InGaAs mHEMT technology with a gate length 35 nm [167] and operate in a frequency range around 240 GHz [169]. A simplified block diagram of the Tx and RX MMIC IQ mixer is shown in Fig. A.6(a,b).

Both the Tx and Rx MMICs consist of the same building blocks. The frequency conversion is implemented by a subharmonic IQ mixer which comprises two mixer transistors (one for I and one for Q modulation) acting as a voltage controlled resistor (resistive mixer). In the subharmonic mixer approach, the LO is located at half of the mm-wave carrier frequency, and mixing is performed with the second harmonic of the LO. The 120 GHz LO signal is split on-chip using a Wilkinson power divider. A balanced mixer concept was employed to obtain a good LO to RF isolation at the Tx MMIC. Therefore, the LO signal in each LO path is split again in two arms and a $\lambda/2$ delay line at the fundamental LO frequency is placed in one of the arms (not shown in Fig. A.6(a,b)). The LO signal is fed to the mixer transistor where the 120 GHz signal cancels itself at the transistor's drain while the mixing products at 240 GHz are interfering constructively [169]. It should be noted that the required 90° phase shift for IQ (de)modulation is not implemented in one of the LO paths as shown in Fig. 2.2. Here, the 90° degree phase shift is realized in the RF path by the Lange coupler which combines the up-converted I and Q signals in the Tx MMIC and splits the received 240 GHz signal in the Rx MMIC, respectively. A three-stage LNA with a small-signal gain of $\approx$30 dB amplifies the Tx MMIC output signals and the received mm-wave signal at the Rx MMIC, respectively.

The Tx and Rx chips have an identical size of 2.5×1 mm^2 and are packaged into split-block waveguide modules using a WR-8 (2.03×1.02 mm^2) waveguide at the LO port and a WR-3 (0.86×0.43 mm^2) waveguide at the RF port. In the packaged module, the Tx achieves a maximum output power of −3.6 dBm when applying a continuous 2 GHz input signal with 0 dBm and a LO power of 7 dBm. The Rx module has a maximum conversion gain of 3.8 dB at 240 GHz, the IQ imbalance stays below 1.9 dB in the frequency range 200…280 GHz, and the NF is about 10 dB [C6]. An optimum signal power of about −10 dBm at the I and Q input ports was experimentally determined for the MMIC IQ Tx module. From Rx sensitivity measurements with 25…40 Gbit/s BPSK signals [C6], an optimum RF input power of −32 dBm was found for the MMIC IQ Rx. Further information on the Tx and Rx MMIC chips and modules can be found in [169][C6][C8].

S-Parameters of the MMIC IQ Mixer Modules

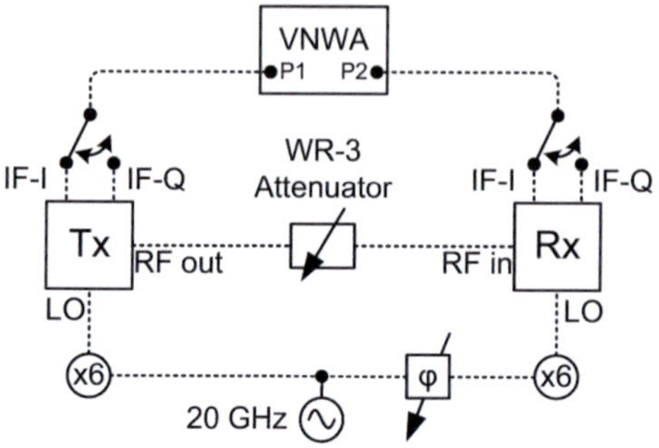

Fig. A.7: Coherent setup for *S*-parameter measurement of the connected Tx and Rx MMIC IQ mixer modules that were introduced in *Section 2.6.1*. Port 1 of the vector network analyzer (VNWA) is either connected to the IF-I or the IF-Q input of the Tx MMIX IQ mixer module while port 2 of the VNWA is connected either to the IF-I or the IF-Q output port of the Rx MMIX IQ mixer module. In this way, all I-Q-Tx-Rx combinations can be measured. The waveguide attenuator connecting the Tx and Rx module is adjusted for the optimum Rx input power.

In [C8], we connected the Tx and Rx IQ mixer modules with a WR-3 waveguide attenuator. A vector network analyzer (VNWA) was used to measure the scattering (*S*) parameters through both modules. Fig. A.7 shows the setup for measuring the *S*-parameters of the combined Tx and Rx MMIC IQ mixer modules.

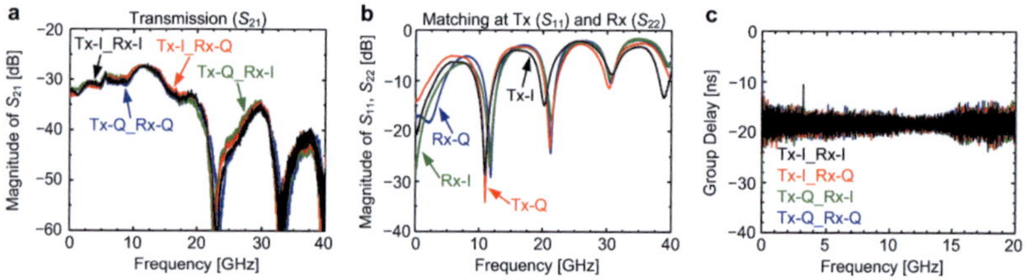

Fig. A.8: *S*-parameter measurement results: (a) transmission (S_{21}) through both Tx and Rx modules, (b) matching at the transmitter (S_{11}) and receiver (S_{22}) IF ports, and the group delay calculated from the phase information of S_{21}.

Fig. A.8 shows the results of the *S*-parameter measurements. In Fig. A.8(a), a (baseband) bandwidth of 20 GHz is observed from the S_{21} measurements of the combined Tx and Rx modules. A baseband bandwidth of 20 GHz means that an RF bandwidth of 40 GHz is available around the carrier frequency of

240 GHz. Fig. A.8(b) shows that the matching at the IF ports of the Tx and Rx modules, is in the range of −5 dB…−3 dB. This moderate matching results from the mixer design which was optimized for a large bandwidth. Furthermore, the IF ports of the Tx and Rx modules have a characteristic impedance of 65 Ω which leads to a mismatch in the 50 Ω measurement environment. The group delay shown in Fig. A.8(c) was calculated from the phase information of the S_{21} measurements. The group delay varies less than 5 ns in the frequency range 0…20 GHz.

Limitations of the MMIC Tx IQ Mixer with Multi-Level Signals

In [C6] we showed that the MMIC Tx and Rx IQ mixers support BPSK data transmission with up to 40 Gbit/s.

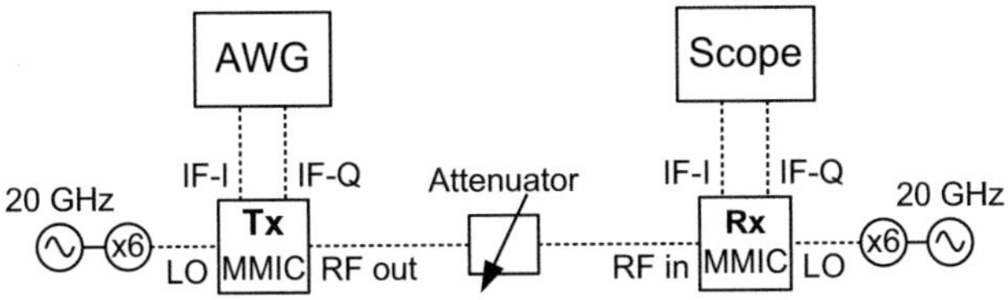

Fig. A.9: Experimental setup for electronic up- and down-conversion of IQ data signals. AWG: arbitrary waveform generator, IF: intermediate frequency, I: inphase, Q: quadrature, LO: local oscillator, RF: radiofrequency, Tx: transmitter, Rx: receiver, MMIC: mm-wave monolithic integrated circuit.

Here, we test the IQ mixers with a 10 GBd QPSK (20 Gbit/s), a 10 GBd 8PSK (30 Gbit/s) [C8], and a 5 GBd 16QAM (20 Gbit/s) signal. The experimental setup is given in Fig. A.9. The IQ baseband data signals are generated with an arbitrary waveform generator (AWG). The average signal input power to the I and Q port of the Tx MMIC module is −9 dBm. The Tx and Rx MMIC modules are connected with a variable waveguide attenuator and the received power at the Rx module is adjusted to −35 dBm. The down-converted IQ signal is recorded with a real-time oscilloscope at a sample rate of 80 GSa/s. Carrier recovery and demodulation of the signal is performed offline with DSP using the Agilent vector signal analysis (VSA) software.

Fig. A.10(a,b) show the constellation diagrams of the received QPSK and 8PSK signals. For both signals, the received symbols are clearly separated and a very good signal quality is observed.

Fig. A.10(c) shows the constellation diagram of the received 16QAM signal. The colored dots in Fig. A.10(c) indicate the ideal constellation points of the 16QAM signal. The three amplitude levels are represented by the colors red (smallest amplitude), green (medium amplitude), and blue (largest amplitude). The 16QAM signal has three amplitudes levels compared to the QPSK and 8PSK signals which both only have one amplitude level. From Fig. A.10(c) it can be seen that the received 16QAM symbols are not correctly demodulated. The different amplitude levels are compressed and cannot be distinguished from each other. This result is explained with the saturation of the LNA in the MMIC Tx module.

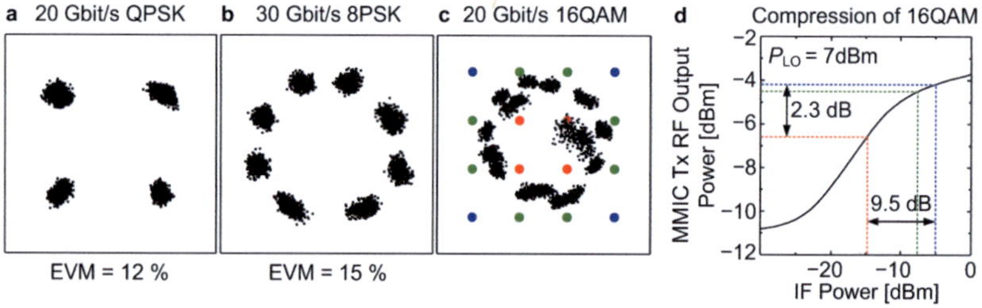

Fig. A.10: Constellation diagrams of the received QPSK (a), 8PSK (b), and 16QAM (c) data signals after electronic up-conversion to 240 GHz and subsequent electronic down-conversion. The constellation diagrams of the QPSK and 8PSK signals with only one amplitude level show a good signal quality. The received symbols of the 16QAM signal with three different amplitude levels are not distinguishable. (d) The limited dynamic range of the Tx mixer compresses the amplitude levels of the 16QAM signal.

Fig. A.10(d) shows the MMIC Tx module's RF output power versus input power characteristic for a LO power of +7 dBm. The colored vertical lines indicate the three different 16QAM input power levels. The average 16QAM input power level of −9 dBm corresponds to the power level of the medium 16QAM amplitude (green color). The ratio between the largest (blue) and smallest (red) 16QAM power levels is 9.5 dB. However, due to the LNA's saturation this ratio gets compressed to only 2.3 dB at the output of Tx MMIC IQ mixer (horizontal lines). By applying a smaller input power level, the mixer could be operated in a more linear region. However, experiments revealed that the influence of noise quickly becomes dominant at lower input power levels.

The limited linear dynamic range of the Tx IQ mixer so far restricts the operation to PSK signals with only one amplitude level. Probably a redesign of the MMIC Tx IQ mixer can increase the linear dynamic range and enable the operation with multi-level IQ signals.

A.5 Uni-Travelling Carrier Photodiode

When using optical heterodyning for mm-wave signal generation, the photodetector is a key device. Not only must it provide a sufficiently large bandwidth to generate the desired wireless carrier frequency, it also should be capable of handling high optical input powers to generate RF signals with acceptable output power. Specially designed UTC-PDs feature a very large bandwidth (> 300 GHz), a large linear dynamic range, and can be operated at high optical input powers. In *Section A.5.1*, the operation principle of a UTC-PD is explained. In *Section A.5.2*, we characterize the UTC-PD that is used for the experiments of this thesis.

A.5.1 Operation Principle of a UTC-PD

The following descriptions and the comparison between the operation principle of a pin PD and a UTC-PD closely follow [42, 194-197].

In Fig. A.11(a,b) the band diagram of a conventional pin PD and a UTC-PD are shown. The pin PD is operated under reverse bias, and the light is absorbed in the depleted intrinsic InGaAs layer where electron-hole pairs are generated. Both electrons and holes contribute to the PD response. Due to the applied electric field, electrons and holes drift to the n-contact and p-contact, respectively. On the one hand, the speed of the pin PD is determined by the carrier transit time through the absorption layer, on the other hand the RC time constant limits the speed of the device. The carrier transit-time limited bandwidth is dominated by the holes, since their propagation velocity is 6…10 times slower than that of electrons. The RC limited bandwidth $f_{3\text{dB}} \propto 1/RC = w_A/(\varepsilon_0\varepsilon_r AR)$ is mainly determined by the thickness w_A of the absorbing layer with dielectric constant $\varepsilon_0\varepsilon_r$ and the device area A. Reducing w_A increases the transit-time limited bandwidth, however only until the RC limited bandwidth of the pin PD starts to dominate.

In contrast to the pin PD, the response of a UTC-PD is only determined by the (fast) electrons. This is achieved by a special design of the active region of the UTC-PD which consists of two layers. One layer is the neutral (undepleted) narrow-bandgap light absorption layer (p-type InGaAs, thickness w_A). The second layer is the undoped or slightly n-type doped (depleted) wide-bandgap carrier collection layer (InP, thickness w_C). Electron-hole pairs are solely generated in the absorption layer. Another wide-bandgap p-type doped layer is placed between the absorption layer and the p-contact to prevent electron diffusion into the p-contact and to force an effective unidirectional motion of the photo-generated electrons towards the collection layer.

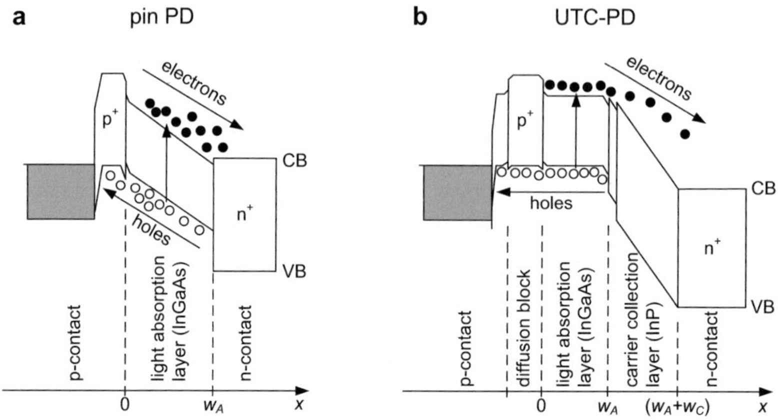

Fig. A.11: Band diagram of (a) conventional pin PD and (b) UTC-PD (drawn after [42]). CB: conduction band, VB: valence band.

In the absorption layer, the photo-generated electrons are the minority carriers. The transport of the holes as majority carriers is a collective motion and the hole transport is limited by the dielectric *relaxation time* (very short, about 1 ps). Thus the response of the whole structure is determined only by the electron transport (in a pin PD both electrons and holes contribute to the response). In the absorption layer, the electron *diffusion* time determines the operation speed. High electron mobilities are necessary to minimize the absorption layer travelling time τ_A which is the time that electrons need to diffuse from the absorption layer into the collection layer. For a relatively thin absorption layer, the electron diffusion velocity in InGaAs can be very large.

By applying a reverse bias voltage to the UTC-PD, electron velocity overshoot is achieved in the collection layer and a collection layer traveling time $\tau_C << \tau_A$ results for a typical UTC-PD structure with $w_A \approx w_C$. The electron overshoot velocity is about one order of magnitude greater than the hole saturation velocity. This leads to a much higher carrier-transport limited bandwidth for a UTC-PD compared to a pin PD, where the carrier transit time of the holes determines the speed.

As mentioned, the overall UTC-PD operation speed is determined by the electron diffusion time in the undepleted absorption layer. This is the reason why for a large absorption layer thickness w_A, the UTC-PD bandwidth is normally smaller than that of a pin PD [42]. When decreasing w_A, the 3 dB bandwidth increases with the inverse square of w_A for the UTC-PD (due to electron diffusion transport) and with the inverse of w_A for the pin PD (due to drift transport of holes and electrons). However, for very small w_A, the bandwidth of the pin PD decreases again because the RC time constant increases. In contrast, the short transit time in the collection layer of a UTC-PD allows significantly reducing the absorption layer thickness w_A without being limited by the RC time constant.

For large optical input powers, the band profile in a pin PD is modulated (band bending), since the photogenerated carriers are stored in the absorption layer. The decreased electric field drastically reduces the carrier velocity, enhances the charge storage, and thus the output current saturates [194].

In a UTC-PD, high optical input powers and high output saturation currents are possible at high-frequency operation, due to a small space charge effect in the collection layer. The space charge only consists of electrons whose velocity at overshoot is much higher than that of holes even for the decreased electric field. Therefore, the output current does not saturate until the current density becomes an order of magnitude larger than that of a pin PD [194].

Several methods exist to optimize a UTC-PD with respect to its responsivity, bandwidth, high optical input power handling capability, and high RF output power. Furthermore, impedance matching, coupling to an antenna, and packaging are import aspects which need to be considered when building a UTC-PD photomixer module. For a detailed description addressing these issues, the reader is referred to [165, 194].

A.5.2 Characterization of UTC-PD

The content of this Subsection has been published in the supplementary information of the journal publication [J3]:

> **S. Koenig**, D. Lopez-Diaz, J. Antes, F. Boes, R. Henneberger, A. Leuther, A. Tessmann, R. Schmogrow, D. Hillerkuss, R. Palmer, T. Zwick, C. Koos, W. Freude, O. Ambacher, J. Leuthold, and I. Kallfass, "Wireless sub-THz communication system with high data rate," *Nature Photon.*, vol. 7, no. 12, pp. 977-981, 2013, published online 13 October 2013, doi: 10.1038/nphoton.2013.275. © 2013 NPG.

Measurement Setup

The UTC-PD in our experiments is similar to the one presented in [165]. The measurement setup to characterize the frequency response and the linearity of the UTC-PD photomixer module is depicted in Fig. A.12. Two continuous wave (cw) external cavity lasers (ECL, Agilent N7714A) with frequencies f_a and f_b and identical output power are combined using a 3 dB polarization maintaining (PM) coupler. The combined signal with a total optical power P_{opt} is then superimposed on the UTC-PD where the RF carrier with frequency $|f_a - f_b|$ is generated by optical heterodyning. The output power P_{RF} of the generated RF carrier is then measured using an Erickson PM4 calorimeter. The responsivity of the UTC-PD (average UTC-PD photocurrent divided by total optical input power) was determined to be 0.26 A/W.

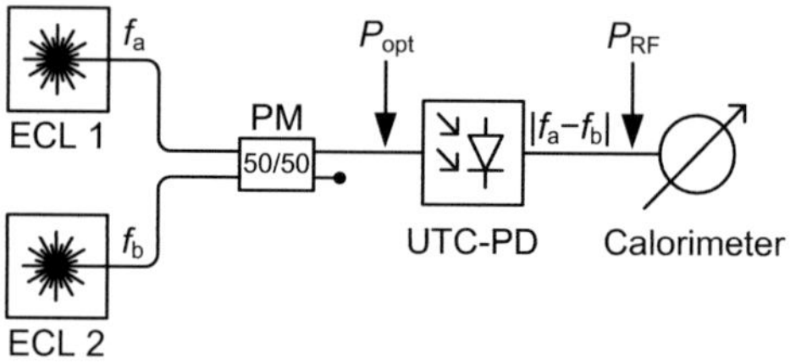

Fig. A.12: Measurement setup for determining the RF frequency response and the linearity of the UTC-PD photomixing module. Both continuous wave external cavity lasers (ECL) have same power. The power P_{RF} of the generated RF carrier is measured with a calorimeter. PM=polarization maintaining. Reprinted from supplementary information of [J3] © 2013 NPG.

UTC-PD Output Power, Frequency Response, and Linearity

First we measure the modulus of the frequency response by tuning the laser frequency f_a with respect to the laser frequency f_b. Fig. A.13(a) shows the measured UTC-PD RF output power P_{RF} as a function of the generated frequency $|f_a - f_b|$ for a total optical input power of $P_{opt} = +13$ dBm. For frequencies > 270 GHz the UTC-PD frequency response is relatively flat. However, the range 200…270 GHz shows a slope and ripples, since the UTC-PD and its RF packaging are not optimized for this frequency range. The vertical dashed lines in Fig. A.13b indicate the positions of the carriers f_1, f_2, and f_3 used in our data transmission experiments in *Section 4.2* and *Section 4.3*. Since the electrical Rx module works best in the region 220…260 GHz, we compensate the non-flat UTC-PD frequency response by optical pre-compensation with the programmable filters at the Tx, and by digital equalization at the Rx.

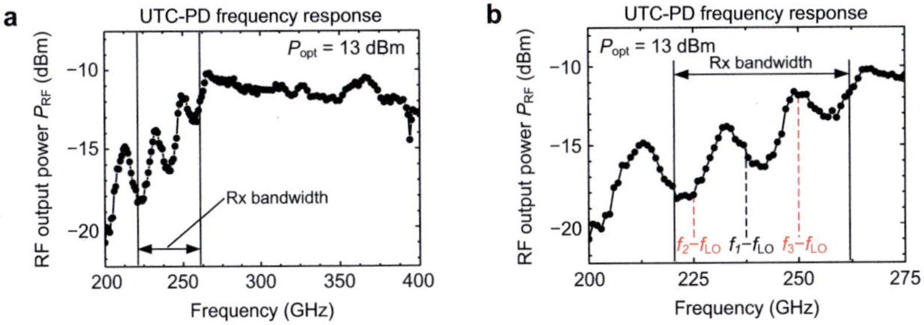

Fig. A.13: UTC-PD frequency response in the range of 200…400 GHz. The non-flat frequency response in the range 200…270 GHz results mainly from the non-optimized RF package in this frequency range. (b) UTC-PD frequency response with slope and ripples in the range 200…270 GHz. The vertical dashed lines indicate the carrier frequencies used in our wireless data transmission experiments. Reprinted from supplementary information of [J3] © 2013 NPG.

For determining the linearity of the UTC-PD, we measure the RF output power P_{RF} as a function of the total optical input power P_{opt} (ECL1 and ECL2 with equal optical power) for a fixed RF frequency $|f_a - f_b|$.

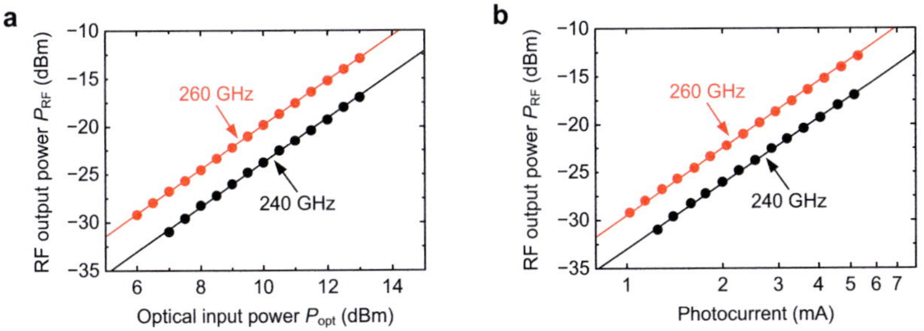

Fig. A.14: Generated RF output power P_{RF} as a function of (a) the total optical input power and (b) the average UTC-PD photocurrent. The UTC-PD shows very good linearity which makes it also suitable for generating multi-level modulation formats such as 16QAM. Reprinted from supplementary information of [J3] © 2013 NPG.

Fig. A.14(a) shows the measurement results for $|f_a - f_b| = 240$ GHz and 260 GHz. A linear relationship between P_{RF} and P_{opt} can be seen irrespective of the generated frequency. At small RF output powers the measurement was limited by the sensitivity (≈ −30 dBm) of the calorimeter. At large power levels, the measurement was limited by the optical output power of the ECLs, and by the damage threshold of the UTC-PD. Very good linearity of P_{RF} is observed in a range of at least 15 dB. Linearity of the UTC-PD is important when generating advanced modulation formats such as 16QAM. Fig. A.14(b) shows the linear relationship between generated RF output P_{RF} power and the average UTC-PD current (logarithmic scale).

A.6 Details on Experimental Setups

A.6.1 SOA Experiments

Fig. A.15 gives details on the setup shown in Fig. 3.2:

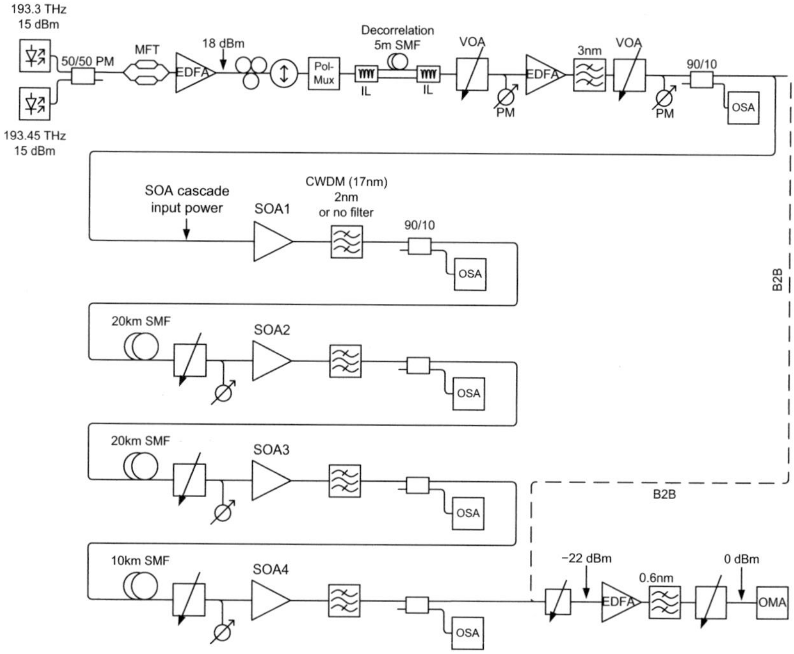

Fig. A.15: SOA cascade with advanced modulation formats. PM: polarization maintaining, MFT: multi-format transmitter, EDFA: erbium doped fiber amplifier, Pol.Mux: polarization multiplexing, SMF: single mode fiber, IL: interleaver, VOA: variable optical attenuator, OSA: optical spectrum analyzer, SOA: semiconductor optical amplifier, CWDM: coarse wavelength division multiplexing filter, B2B: back-to-back, OMA: optical modulation analyzer.

Fig. A.16 gives details on the setup shown in Fig. 3.23(a):

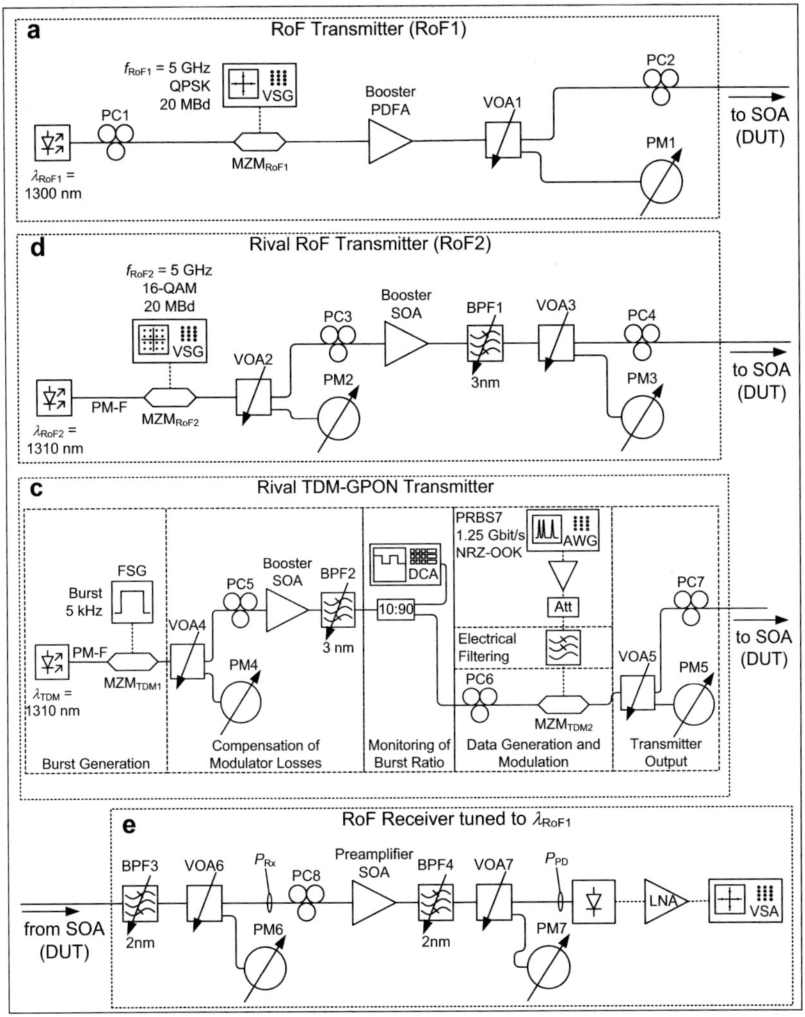

Fig. A.16: Experimental setup for the amplification of a radio-over-fiber (RoF) signal with rival signals. PC: polarization controller, MZM: Mach-Zehnder modulator, VSG: vector signal generator, PFDA: praseodymium-doped fiber amplifier, VOA: variable optical attenuator, PM: power meter, SOA: semiconductor optical amplifier, DUT: device under test, PM-F: polarization-maintaining fiber, BPF: bandpass filter, FSG: function signal generator, DCA: digital communications analyzer, PRBS: pseudorandom bit sequence, NRZ-OOK: non-return-to-zero on-off keying, AWG: arbitrary waveform generator, LNA: low noise amplifier, VSA: vector signal analyzer.

Fig. A.17 gives details on the setup shown in Fig. 3.32(a):

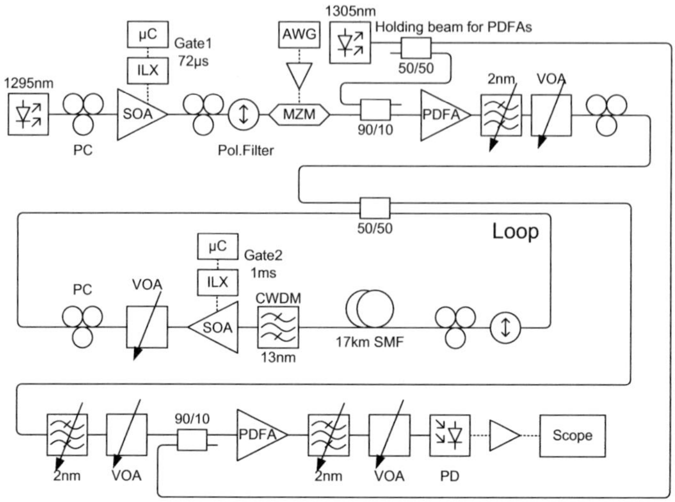

Fig. A.17: Recirculating loop experiment with a semiconductor optical amplifier (SOA). PC: polarization controller, µC: micro-controller, ILX: current source, MZM: Mach-Zehnder modulator, AWG: arbitrary waveform generator, PDFA: praseodymium-doped fiber amplifier, VOA: variable optical attenuator, CWDM: coarse wavelength division multiplexing filter, SMF: single-mode fiber, PD: photodiode, Scope: realtime oscilloscope.

A.6.2 Wireless Link with mm-Wave Photonics

The content of this Subsection has been published in the supplementary information of the journal publication [J3]:

> **S. Koenig**, D. Lopez-Diaz, J. Antes, F. Boes, R. Henneberger, A. Leuther, A. Tessmann, R. Schmogrow, D. Hillerkuss, R. Palmer, T. Zwick, C. Koos, W. Freude, O. Ambacher, J. Leuthold, and I. Kallfass, "Wireless sub-THz communication system with high data rate," *Nature Photon.*, vol. 7, no. 12, pp. 977-981, 2013, published online 13 October 2013, doi: 10.1038/nphoton.2013.275. © 2013 NPG.

Photonic Up-Conversion with UTC-PD: Setup Details

Fig. A.18 (central row) shows a detailed experimental setup of our photonic wireless THz transmitter (Tx) as used in *Section 4.3*. A mode-locked laser (MLL, Time Bandwidth Products ERGO XG) generates a frequency comb with a line spacing of $\Delta f_{\text{MLL}} = 12.5$ GHz. Inset (A) shows the relative MLL power spectrum measured in a resolution bandwidth (RBW) of 1.25 GHz (0.01 nm) and normalized to the strongest carrier. The vertical dashed lines indicate the carriers that are used as local oscillator (LO, green, $f_{\text{LO}} = 193.138$ THz) and optical carriers (black, $f_1 = 193.3755$ THz, red, $f_{2,3} = f_1 \mp \Delta f_{\text{MLL}}$). The frequency spacing between the LO and the central optical carrier is 237.5 GHz. The MLL frequency comb is amplified by an erbium doped fiber amplifier (EDFA).

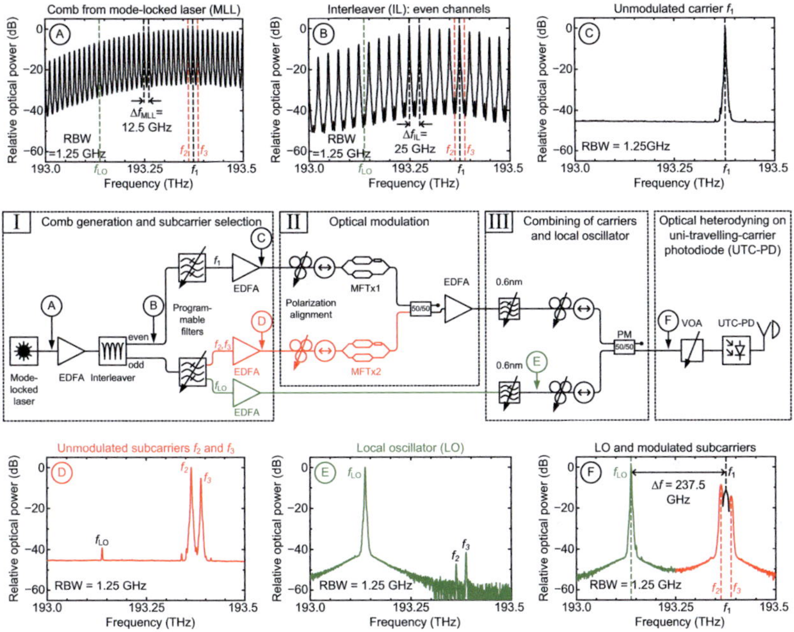

Fig. A.18: Detailed experimental setup of the photonic part of the wireless THz transmitter, where the radiated mm-wave signals are generated by optical heterodyning with a uni-travelling-carrier photodiode (UTC-PD). RBW = resolution bandwidth, EDFA = erbium doped fiber amplifier, MFTx = multi-format transmitter, PM = polarisation maintaining coupler, VOA = variable optical attenuator. Reprinted from supplementary information of [J3] © 2013 NPG.

The subsequent 12.5/25 GHz optical dis-interleaver (Optoplex IL-CABE2S001) splits the frequency comb into even and odd lines and provides a suppression of ≈35 dB of neighboring lines. The optical spectrum at the even output port is shown in inset (B) and includes carrier f_1. The spectrum at the odd output port (not shown) includes the LO at f_{LO} and the carriers f_2 and f_3. All other frequency comb lines must be removed.

To select the desired lines, we use two programmable optical filters (Finisar WaveShaper 4000E), the outputs of which are amplified by three EDFA, Box I. The programmable filter at the even output port of the dis-interleaver removes all comb lines except for carrier f_1, see inset (C). A second waveshaper is connected to the odd output port of the dis-interleaver. The waveshaper distributes frequencies f_2 and f_3 to one of its four output ports (inset (D)), and the LO frequency f_{LO} to another one. Although the chosen ports have the smallest cross-talk, a residual LO contribution suppressed by at least 35 dB compared to carrier f_3 can be seen in inset (D). This LO remainder will be further suppressed by a 0.6 nm bandpass filter that selects the carriers f_2 and f_3 after modulation (Fig. A.18, central row, Box III, upper arm). The LO is filtered with another 0.6 nm bandpass filter, Box III, lower arm. The corresponding spectrum is shown in inset (E). The suppression of the residual carriers f_2 and f_3 is > 40 dB.

The central carrier f_1 and the adjacent carriers $f_{2,3}$ are independently modulated with different IQ data (pseudo random bit sequence length $2^{15}-1$), which are generated by two multi-format transmitters [83] MFTx1 and MFTx2, Box II. In the multi-channel case all three carriers f_1, f_2, f_3 are transmitted, while for single-channel transmission only carrier f_1 is selected (f_2, f_3 are disconnected from the waveshaper). In our experiments we generate the following single-channel signals: 12.5 GBd 16QAM (50 Gbit/s), 25 GBd QPSK (50 Gbit/s), 25 GBd 8QAM (75 Gbit/s), and 25 GBd 16QAM (100 Gbit/s). For multi-channel transmission, we use sinc-like pulse shaping with a raised-cosine spectrum and a roll-off factor $\beta = 0.35$ in all three channels. The symbol rate for carrier f_1 (Ch 1) is 13 GBd. Carriers $f_{2,3}$ (Ch 2, Ch 3) have a symbol rate of 8 GBd. In our experiments we generate two different multi-channel signal settings: If all carriers are modulated with QPSK, this leads to an aggregate data rate of 58 Gbit/s. By using 16QAM in Ch 1 and 8 QAM in Ch 2 and Ch 3, an aggregate data rate of 100 Gbit/s results.

A 3 dB coupler combines the modulated signals from MFTx1 and MFTx2. The combined signal is amplified by an EFDA (overall output power +18 dBm, Box II), and the out-of-band noise is filtered by a 0.6 nm bandpass filter, Box III.

The signals in both arms of Box III are polarization-aligned to maximize the wireless THz signal power generated by the optical heterodyning process at the UTC-PD. A polarization maintaining (PM) 3 dB coupler then combines both signals. The optical spectrum for the multi-channel case is shown in inset (F). Ch 1, Ch 2, and Ch 3 have different signal strengths for compensating the frequency response of the UTC-PD (refer to the UTC-PD section in this document for UTC-PD details).

The total optical input power to the UTC-PD is adjusted by a variable optical attenuator (VOA). We use a PM free-space attenuator. The optical input power to the UTC-PD is set to +14 dBm and generates an average photocurrent of 6.5 mA. The resulting RF output power is −13.5 dBm for a single modulated THz channel at $f_1 - f_{\mathrm{LO}} = 237.5$ GHz.

Fig. A.19 shows a photograph of the last part of the wireless THz transmitter. The fiber which carries the LO and modulated signals is connected to the UTC-PD photomixer module. After photomixing, the THz signal is radiated from a conical horn antenna, which is directly attached to the UTC-PD. The horn antenna has a gain of 22.1 dBi and an aperture with a diameter of 6 mm. The radiated signal is focused by an aspherical plano-convex lens made of high-density polyethylene. The lens diameter is 10.3 cm. The whole setup is stably mounted on a tripod.

Fig. A.20 shows a photograph of the electronic mm-wave receiver (Rx) setup mounted on top of a tripod. The Rx setup consists of a frequency sextupler which drives the LO input of a subharmonic 240 GHz receiver module. Both modules employ active mm-wave monolithic integrated circuits (MMIC). Like on the Tx side, the wireless signal with a center frequency of 237.5 GHz is focused by an aspherical plano-convex lens, and received with a conical horn antenna. The Rx input power can be adjusted by a variable waveguide attenuator. A 19.833 GHz signal from an external synthesizer is sextupled on-chip to 118.75 GHz and then fed to the LO port of the subharmonically pumped MMIC IQ-mixer.

The received RF signal is directly down-converted to the baseband, where the in-phase (I) and quadrature (Q) signals are provided at the respective I and Q output ports. More detailed information about the Rx MMIC IQ-mixer can be found in ref. [169], [C6].

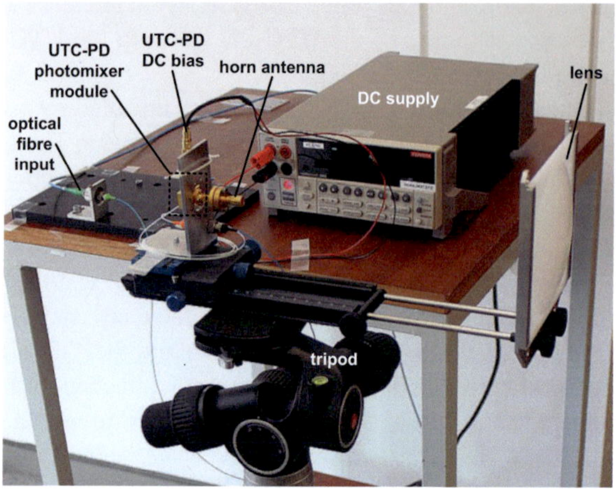

Fig. A.19: Photograph of the UTC-PD photomixer module. The PD is fed via an optical fiber and radiates mm-waves from a beam-focusing combination of conical horn antenna and aspherical plano-convex lens. Reprinted from supplementary information of [J3] © 2013 NPG.

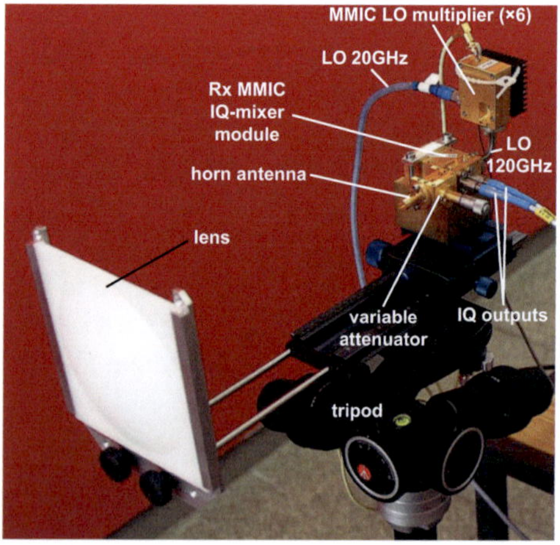

Fig. A.20: Photograph of the electronic mm-wave receiver (Rx) mounted on a tripod. The wireless signal is received by an identical beam focusing lens-antenna combination as used on the Tx side. The input power to the Rx MMIC IQ-mixer module can be adjusted by the variable waveguide attenuator. The received wireless signal with centre frequency 237.5 GHz is directly down-converted to the baseband. Reprinted from supplementary information of [J3] © 2013 NPG.

Estimation of Phase Noise

Since we cannot directly measure the phase noise specifications of the 237.5 GHz signal which is generated by optical heterodyning selected MLL comb lines in the UTC-PD, we analyze the phase noise of RF signals with a carrier frequency of 12.5 GHz, 25 GHz, and 50 GHz. From these results we then extrapolate the phase noise specifications of the 237.5 GHz signal.

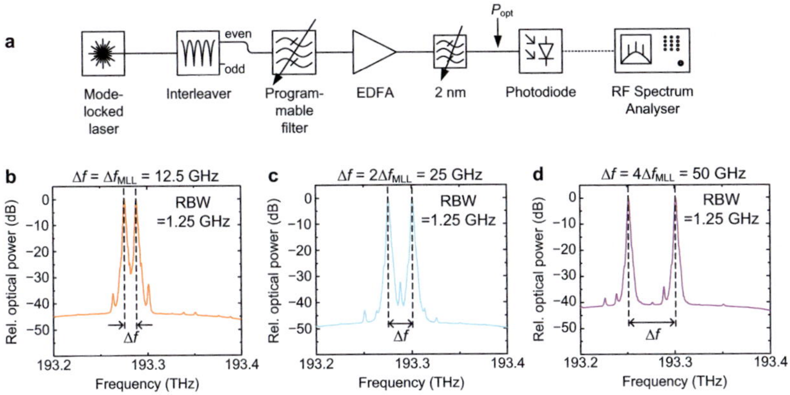

Fig. A.21: Measuring linewidth and phase noise of optically generated radiofrequency (RF) carriers. (a) RF carrier generation by optical heterodyning two selected lines of a mode-locked laser (MLL). (b, c, d) Optical spectrum before the photodiode with a frequency spacing of (a) 12.5 GHz (without interleaver), (b) 25 GHz, and (c) 50 GHz between the two selected MLL lines. EDFA = erbium doped fiber amplifier, RBW = resolution bandwidth. Reprinted from supplementary information of [J3] © 2013 NPG.

Fig. A.21a shows the measurement setup. With the programmable filter (waveshaper) we select two MLL comb lines that have a frequency spacing of $\Delta f = \Delta f_{\text{MLL}}$ = 12.5GHz, $\Delta f = 2\Delta f_{\text{MLL}}$ = 25GHz, or $\Delta f = 4\Delta f_{\text{MLL}}$ = 50GHz, respectively. For a frequency spacing of 25 GHz and 50 GHz, we use a 12.5/25 GHz optical interleaver in front of the waveshaper to achieve a good suppression of neighboring comb lines. The waveshaper distributes both selected MLL lines to the same output port, i.e., both are coupled into the same fiber. This simplifies the setup, because neither polarization alignment of the different fiber paths nor a fiber coupler is needed in front of the PD. After the waveshaper, the two lines are amplified by an EDFA and then filtered with a 2 nm bandpass filter. Both lines have identical powers and the total optical input power P_{opt} to the subsequent pin-PD (u2t 3120R, 70 GHz electrical 3 dB bandwidth) is kept constant at +8 dBm. Fig. A.21(b,c,d) show the optical spectra at the PD input for the three different frequency spacings. After photomixing, we analyze the phase noise of the generated RF carrier with an electrical spectrum analyzer (Agilent E4448A with phase noise measurement option). Due the bandwidth limitation of the spectrum analyzer (3 Hz to 50 GHz) and because of the fixed frequency spacing of the MLL, the largest generated carrier frequency is 50 GHz. We wanted to use only one interleaver output and one fiber setting to come as close as possible to our transmission measurement setup, so we had only the choice of the three frequencies 12.5 GHz, 25 GHz, and 50 GHz. We could have connected the programmable filter (waveshaper) directly to the MLL, however, then the resulting frequency selectivity would have been about 10 dB worse than with an interleaver (compare Fig. A.18a,b and Fig. A.21b,d), and this would also cast some doubt on the measurements.

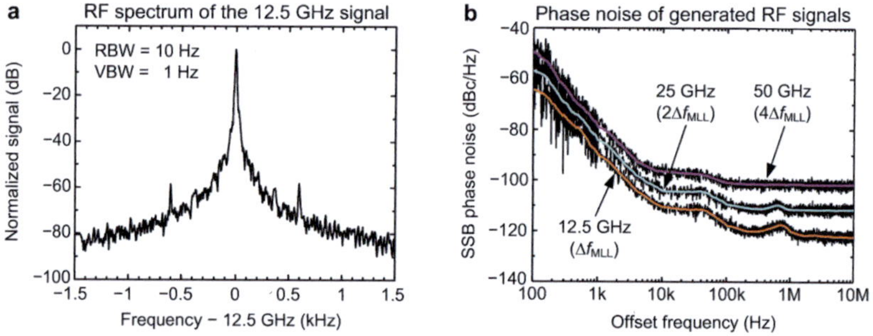

Fig. A.22: RF spectra and phase noise. (a) Spectrum of the generated 12.5 GHz RF carrier acquired over a 3 kHz span with a resolution bandwidth (RBW) of 10 Hz and a video bandwidth (VBW) of 1 Hz. The spectrum is centred at the carrier frequency. The origin of the sidelobe peaks at about 0.6 kHz carrier offset is unknown. (b) Single sideband (SSB) phase noise measurement of the 12.5 GHz, 25 GHz, and 50 GHz carriers. The solid lines result from averaging the measured data. Reprinted from supplementary information of [J3] © 2013 NPG.

Fig. A.22a shows the measured spectrum of the generated 12.5 GHz carrier over a span of 3 kHz and centered on the carrier frequency. The trace was obtained with a RBW of 10 Hz and a video bandwidth (VBW) of 1 Hz. A measurement with an even smaller RBW and a smaller span was impractical, since the generated carrier frequency was slowly drifting by some hundreds of Hz. We attribute the drifting to thermal fluctuations in the MLL. From the trace in Fig. A.22a we infer that the 12.5 GHz carrier has a linewidth < 100 Hz. As mentioned before, the linewidth of a single comb line has a width of about 1 kHz which was measured in another heterodyne experiment [170]. Since the phase noise of the different MLL lines is correlated, the linewidth of the generated RF carrier is smaller than the linewidth of a single MLL line. The origin of the sidelobe peaks at about 0.6 kHz carrier offset could not be identified.

Fig. A.22b shows the measured single sideband (SSB) phase noise (dBc/Hz) of the generated 12.5 GHz, 25 GHz, and 50 GHz carrier as a function of the offset frequency to the carrier (colored lines represent averaged data). For an offset frequency smaller than 10 kHz, the phase noise decays with frequency with a slope of 20 dB per decade. Above a frequency offset of 10 kHz the noise spectrum is virtually white.

The SSB phase noise performance (averaged data) at different offset frequencies is summarized in the first three rows of Table S1. The increase of the phase noise when doubling the carrier frequency from 12.5 GHz to 25 GHz and from 25 GHz to 50 GHz is indicated in parentheses in the second and third row, respectively. From these two values in each column at the offset frequencies 100 Hz, 1 kHz, 10 kHz, 100 kHz, 1 MHz, and 10 MHz we calculate an average increase Δ (per octave) of the SSB phase noise of 7.35 dB, 5.5 dB, 7 dB, 8.45 dB, 9.25 dB, and 10.35 dB. The frequency 237.5 GHz can be written as 19×12.5 GHz $= 2^{4.25}\times12.5$ GHz, i.e., it is scaled up from 12.5 GHz by 4.25 octaves. For the optically generated 237.5 GHz carrier, we therefore estimate the SSB phase noise at each offset frequency by adding $4.25\times\Delta$ to the measured SSB phase noise value at 12.5 GHz. The such extrapolated phase noise values for an exemplary carrier frequency 237.5 GHz are given in the first row of Table S2.

In the last row of Table S1, we give the SSB phase noise values (typical data as specified in the data sheet) of a 20 GHz carrier generated by the same synthesizer (Wiltron 68087B) that was employed for the LO generation at the Rx. At an offset frequency of 100 Hz, the synthesizer exhibits a lower phase noise than the 25 GHz carrier generated by photomixing. For larger offset frequencies, the phase noise of the optically generated RF carrier is lower.

The LO at the Rx is generated by multiplying the carrier frequency from the external synthesizer operating around 20 GHz by a factor $N = 12$. We assume that in this multiplication process the SSB phase noise scales with $20\log_{10} N$. The estimated values for the SSB phase noise of the Rx LO carrier are given in Table S2.

SSB phase noise (dBc/Hz)		Offset frequency					
		100 Hz	1 kHz	10 kHz	100 kHz	1 MHz	10 MHz
Carrier frequency f_{RF}	**12.5 GHz**	−64.3	−89.8	−110.5	−117.8	−120	−122.4
	25 GHz	−56.9 (+7.4 dB)	−84.1 (5.7 dB)	−104.1 (6.4 dB)	−109.5 (8.3 dB)	−111.4 (8.6 dB)	−111.5 (10.9 dB)
	50 GHz	−49.6 (+7.3 dB)	−78.8 (5.3 dB)	−96.5 (7.6 dB)	−100.9 (8.6 dB)	−101.5 (9.9 dB)	−101.7 (9.8 dB)
	20 GHz (synthesizer)	−66	−78	−78	−100	n/a	n/a

Table A.2 Measured single sideband (SSB) phase noise (dBc/Hz) specifications. The values in parentheses indicate the phase noise increase per octave, i.e., for column 100 Hz and row 25 GHz, +7.4 dB = −56.9 dB − (−64.3 dB).

SSB phase noise (dBc/Hz)	Offset frequency					
	100 Hz	1 kHz	10 kHz	100 kHz	1 MHz	10 MHz
237.5 GHz (Tx: heterodyning MLL lines)	−33.1	−66.4	−80.8	−81.9	−80.7	−78.4
237.5 GHz (Rx LO: derived from synthesizer)	−44.4	−56.4	−56.4	−78.4	n/a	n/a

Table A.3 Extrapolated single sideband (SSB) phase noise (dBc/Hz) at 237.5 GHz.

Photonic Up-Conversion with UTC-PD: Back-to-back Setup

The back-to-back (B2B) signal quality for both the single-channel (*Section 4.2*) and multi-channel operation (*Section 4.3*) is measured by directly connecting the UTC-PD photomixer module to the variable waveguide attenuator of the electronic Rx as depicted in Fig. A.23. After down-conversion to the baseband, the signal's I and Q components are amplified by a pair of phase matched RF amplifiers. An analog-to-digital converter (ADC) with a sampling rate of 80 GSa/s allows further offline digital signal processing (DSP) including channel equalization, carrier recovery, and demodulation.

Fig. A.24 shows a photograph of the B2B measurement setup for photonic up-conversion.

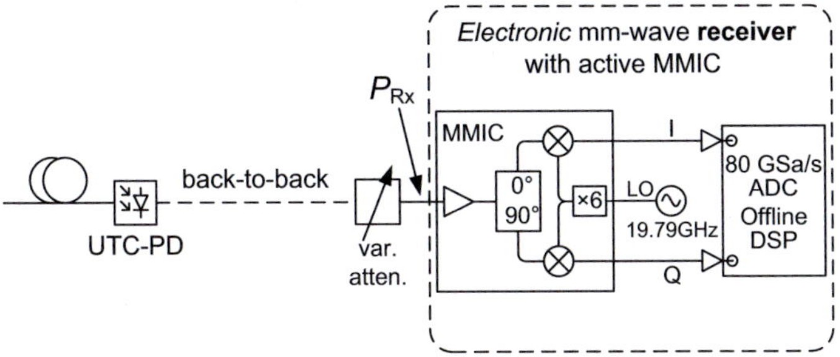

Fig. A.23: Back-to-back measurement setup for photonic up-conversion. The UTC-PD photomixer module is directly connected to the variable attenuator. Reprinted from supplementary information of [J3] © 2013 NPG.

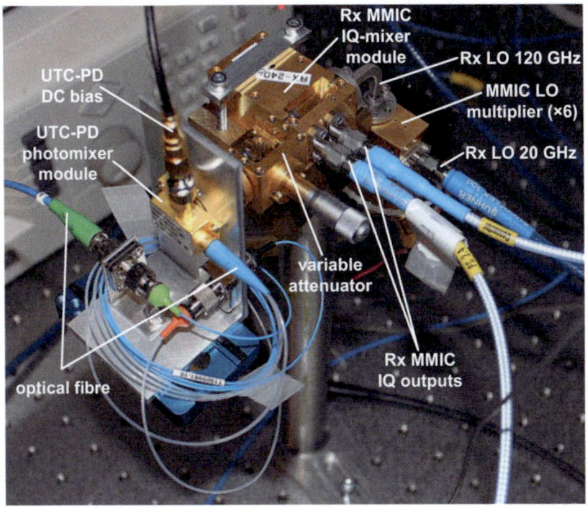

Fig. A.24: Photograph of the back-to-back measurement setup for photonic up-conversion. Reprinted from supplementary information of [J3] © 2013 NPG.

BER, EVM and FEC

In our experiments, we digitized and recorded the received data with a real-time oscilloscope (Agilent DSO-X-93204A) for offline digital signal processing and signal quality evaluation. The recording length was always 80 µs. For a 25 GBd single-channel signal, a recording length of 80 µs corresponds to a total of 2×10^6 symbols, that is, we evaluated ~4×10^6 received bits for the 50 Gbit/s QPSK signal, 6×10^6 bits for the 75 Gbit/s 8PSK signal and 8×10^6 bits for the 100 Gbit/s 16QAM signal.

A widely accepted quality metric for a data signal is the BER. To determine the BER, a known bit sequence must be transmitted. In our experiments, we transmitted a pseudo-random bit sequence of length $2^{15}-1$. At the receiver, we compared the recorded bit sequence with the originally sent bit sequence and counted the number of bit errors. This number divided by the number of compared bits is then an estimate for the BER. However, if the signal quality is high, that is, if the BER is small, a significant amount of time (a very large number of recordings) is needed to count enough errors to determine a reliable BER value.

In our case, the quality of the received QPSK signals was so good that too few or no errors were counted within the recording length of 80 µs. For all other signal formats, enough errors were counted within the recording length, and the BER could be computed. For low-error rate QPSK signals we measured the EVM instead of the BER. The EVM is a standard metric for signal quality in wireless and wireline communication systems. The EVM describes the effective distance of the receive1d complex symbols from their ideal positions in the constellation diagram. Under the assumptions of additive white Gaussian noise, data-aided reception and quadratically arranged xQAM constellations, with $\log_2 x$ being an even number (e.g., $x = 4$ for QPSK), the BER can be estimated from measured EVM values. The horizontal solid lines in Fig. 4.14a and Fig. 4.15a indicate corresponding calculated BER values of 1×10^{-9}.

The horizontal dashed lines in Fig. 4.14 and Fig. 4.15 indicate a BER of 4.5×10^{-3}. A raw BER of 4.5×10^{-3} is the threshold for error-free transmission when using second-generation hard-decision FEC with 7% overhead [171]. If the FEC works properly, the corrected output BER becomes $< 1\times10^{-15}$.

We mentioned in the detailed setup description of Fig. A.18 that we use a 3 dB fiber coupler to combine the modulated signal(s) and the LO before photomixing. As described by Hirata and colleagues [198], variations in temperature might cause fluctuations in the optical path length difference between the signal path and the LO path. This in turn could result in phase fluctuations of the generated terahertz signal on a timescale of several milliseconds to seconds. Thus, within the recording length of 80 µs, we did not see any significant influence of this phase effect on signal quality. However, error bursts would be an issue in practical systems. On the other hand, Hirata and colleagues [198] have shown that error bursts can be avoided if photonic integrated structures and combiners are used instead of discrete fiber-based components.

Constants

$$h = 6.62606957 \times 10^{-34}\ \mathrm{Ws}^2 \tag{A.78}$$

$$e = 1.602176565 \times 10^{-19}\ \mathrm{As} \tag{A.79}$$

$$k = 1.3806488 \times 10^{-23}\ \mathrm{J/K} \tag{A.80}$$

$$\varepsilon_0 = 8.854188 \times 10^{-12}\ \frac{\mathrm{As}}{\mathrm{Vm}} \tag{A.81}$$

$$\mu_0 = 1.256637 \times 10^{-6}\ \frac{\mathrm{Vs}}{\mathrm{Am}} \tag{A.82}$$

$$c_0 = \frac{1}{\sqrt{\varepsilon_0 \mu_0}} = 299792458\ \frac{\mathrm{m}}{\mathrm{s}} \tag{A.83}$$

$$Z_0 = \sqrt{\frac{\mu_0}{\varepsilon_0}} = 376.7303\ \Omega \tag{A.84}$$

Glossary

Acronyms

2G	Second generation (mobile communication technology)
3D	Three dimensional
3G	Third generation (mobile communication technology)
4G	Fourth generation (mobile communication technology)
3R	Reamplification, reshaping, retiming
ADM	Add-drop multiplexer
ADSL	Asymmetric digital subscriber line
AM	Amplitude modulation
ASE	Amplified spontaneous emission
ASK	Amplitude-shift keying (format)
Att.	Attenuator
AWG	Arbitrary waveform generator
AWGN	Additive white Gaussian noise
B2B	Back-to-back
BER	Bit-error ratio
BPON	Broadband PON, broadband passive optical network
BR	Burst ratio
BSC	Base station controller
BtB	Back-to-back
BTS	Base transceiver station
CAGR	Compound annual growth rate
CAPEX	Capital expenditures
CB	Conduction band
CDMA	Code division multiple access
CO	Central office
CoMP	Coordinated multi point
CPE	Customer premise equipment
CW	Continuous-wave
CWDM	Coarse wavelength-division-multiplexing
DCA	Digital communications analyzer
DCF	Dispersion compensating fiber
DD	Direct detection
DP	Dual polarization
DQPSK	Differential quadrature phase-shift keying (format)

DSL	Digital-subscriber line
DSP	Digital signal processing
DUT	Device under test
DWDM	Dense wavelength-division-multiplexing
DXC	Digital-cross connect
EAM	Electro-absorption modulator
ECL	External cavity laser
EDFA	Erbium-doped fiber amplifier
EDGE	Enhanced data rates for GSM
EPDR	Electrical power dynamic range
EPON	Ethernet PON, ethernet passive optical network
ER	Extinction ratio
EVM	Error vector magnitude
FEC	Forward error correction
FSO	Free-space optical
FtF	Fiber-to-Fiber
FTTB	Fiber-to-the-Building
FTTC	Fiber-to-the-Curb
FTTH	Fiber-to-the-Home
FWHM	Full width at half maximum
FWM	Four-wave mixing
GPON	Gigabit PON, gigabit passive optical network
GSM	Global system for mobile communications
HDTV	High-definition television
HSPA	High speed packet access
I	Inphase
IM	Intensity modulation
IP	Internet protocol
IPDR	Input power dynamic range (optical)
LAN	Local area network
LCA	Lightwave component analyzer
LNA	Low noise amplifier
LO	Local oscillator
LTE	Long term evolution
MAC	Media access control (protocol)
MAN	Metropolitan-area network
MIMO	Multiple-input multiple-output
MLL	Mode-locked laser
MMIC	Millimeter-wave monolithic integrated circuit
MZM	Mach-Zehnder modulator

NF	Noise figure
NG-PON	Next-generation PON, next-generation passive optical network
NRZ	Non-return-to-zero
OADM	Optical add-drop multiplexer
OBPF	Optical band-pass filter
OEO	Optical-to-electrical-to-optical (conversion)
OFDM	Optical orthogonal frequency-division multiplexing
OLT	Optical line terminal, sometimes also used with: termination
OMA	Optical modulation analyzer
ONT	Optical-network termination
OOK	On-off-keying
OPEX	Operational expenditures
OPDR	Optical power dynamic range (also known as IPDR)
OSA	Optical spectrum analyzer
OSNR	Optical signal to noise ratio
OXC	Optical-cross connect
P2P	Point-to-point
P2MP	Point-to-multipoint
PC	Polarization controller
PD	Photodiode
PDFA	Praseodymium-doped fiber amplifier
PDG	Polarization dependent gain
PON	Passive optical network
PM	Power meter
PMF	Polarization-maintaining fiber
PRBS	Pseudo-random bit sequence
PSK	Phase-shift keying (format)
Q	Quadrature
QAM	Quadrature-amplitude modulation
QD	Quantum dot
QD SOA	Quantum-dot semiconductor optical amplifier
QW	Quantum well
RE	Reach extender
rms	root mean square
RAN	Radio access network
RBW	Resolution bandwidth
RF	Radiofrequency
RL	Recirculating loop

RNC	Radio network controller
ROADM	Reconfigurable optical add-drop multiplexer
RoF	Radio-over-Fiber
RSOA	Reflective-semiconductor optical amplifier
Rx	Receiver
SBD	Schottky barrier diode
SISO	Single-input single output
SMF	Single-mode fiber
SNR	Signal-to-noise ratio
SOA	Semiconductor optical amplifier
SP	Single polarization
sp	spontaneous (emission)
SRH	Shockly-Rhead-Hall (recombination)
TDM	Time division multiplexing
TP	Twisted pair
TV	Television
Tx	Transmitter
UMTS	Universal mobile telecommunications system
UTC	Uni-travelling carrier
VB	Valence band
VDSL	Very high-capacity digital subscriber line
VNWA	Vector network analyzer
VOA	Variable optical attenuator
VoD	Video-on-demand
VoIP	Voice-over-IP
WAN	Wide-area network
WDM	Wavelength-division-multiplexing
WiMAX	Worldwide interoperability for microwave access
WSS	Wavelength-selective switch
XGM	Cross-gain modulation
XPM	Cross-phase modulation

References

[1] Cisco Systems Inc., "Cisco Visual Networking Index: Forecast and Methodology, 2011-2016," 2012.

[2] Cisco Systems Inc., "Cisco Visual Networking Index: Global Mobile Data Traffic Forecast Update, 2012–2017," 2013.

[3] R. Bonk, "Linear and Nonlinear Semiconductor Optical Amplifiers for Next-Generation Optical Networks," PhD Thesis, Karlsruhe Institute of Technology (KIT), Karlsruhe Series in Photonics & Communications vol. 8, 2013.

[4] M. Sexton, and A. Reid, *Transmission Networking: SONET and the Synchronous Digital Hierarchy*, Boston: Artech House, 1992.

[5] J. Leuthold, "Optische Kommunikationssysteme," Lecture notes, Institute of Photonics and Quantum Electronics, Karlsruhe Institute of Technology (KIT), 2009.

[6] J. Prat, *Next-generation FTTH passive optical networks: research towards unlimited bandwidth access*, Springer, 2008.

[7] B. Mukherjee, *Broadband access networks*, Dordrecht: Springer, 2009.

[8] P. Vorreau, "An optical grooming switch for high-speed traffic aggregation," PhD Thesis, Karlsruhe Institute of Technology (KIT), Karlsruhe Series in Photonics & Communications vol. 6, 2009.

[9] T. Koonen, "Fiber to the Home/Fiber to the Premises: What, Where, and When?," *Proc. IEEE,* vol. 94, no. 5, pp. 911-934, 2006.

[10] I. Kaminov, T. Li, and A. E. Willner, *Optical fiber telecommunications V, B: Systems and networks*, Amsterdam: Elsevier, 2008.

[11] R. P. Davey, D. B. Grossman, M. Rasztovits-Wiech, D. Payne, D. Nesset, A. E. Kelly, A. Rafel, S. Appathurai, and Y. Sheng-Hui, "Long-Reach Passive Optical Networks," *J. Lightw. Technol.,* vol. 27, no. 3, pp. 273-291, 2009.

[12] N. Cvijetic, D. Qian, J. Hu, and T. Wang, "Orthogonal frequency division multiple access PON (OFDMA-PON) for colorless upstream

transmission beyond 10 Gb/s," *IEEE J. Sel. Areas Commun.,* vol. 28, no. 6, pp. 781-790, 2010.

[13] N. Cvijetic, "OFDM for Next-Generation Optical Access Networks," *J. Lightw. Technol.,* vol. 30, no. 4, pp. 384-398, 2012.

[14] P. Schindler, R. M. Schmogrow, D. Hillerkuss, M. Nazarathy, S. Ben-Ezra, C. Koos, W. Freude, and J. Leuthold, "Remote Heterodyne Reception of OFDM-QPSK as Downlink-Solution for Future Access Networks," *Access Networks and In-house Communications (ANIC)*, Colorado Springs (CO), USA, June 2012, paper AW4A.3.

[15] H. G. Krimmel, T. Pfeiffer, B. Deppisch, and L. Jentsch, "Hybrid electro-optical feedback gain-stabilized EDFAs for long-reach wavelength-multiplexed passive optical networks," *European Conference and Exhibition on Optical Communication (ECOC)*, Vienna, Austria, September 2009.

[16] H. Rohde, S. Smolorz, J. S. Wey, and E. Gottwald, "Coherent optical access networks," *Optical Fiber Communication Conference (OFC)*, Los Angeles (CA), USA, March 2011, paper OTuB1.

[17] Rysavy Research LLC., "Mobile Broadband Explosion - The 3GPP Wireless Evolution," *http://www.rysavy.com/papers.html*, 2012.

[18] A. Ghosh, R. Ratasuk, B. Mondal, N. Mangalvedhe, and T. Thomas, "LTE-advanced: next-generation wireless broadband technology [Invited Paper]," *IEEE Wireless Communications,* vol. 17, no. 3, pp. 10-22, 2010.

[19] R. Chundury, "Mobile broadband backhaul: Addressing the challenge," *Ericsson Review,* vol. 3, pp. 4-9, 2008.

[20] J. Hansryd, and P.-E. Eriksson, "High-speed mobile backhaul demonstrators," *Ericsson Review,* vol. 2, pp. 10-16, 2009.

[21] T-Mobile. "T-Mobile's Backhaul Strategy Key to a Competitive 4G Experience," http://www.tmonews.com/2012/08/t-mobile-says-backhaul-strategy-key-to-a-competitive-4g-experience/.

[22] M. Popov, "The convergence of wired and wireless services delivery in access and home networks," *Optical Fiber Communication Conference (OFC)*, San Diego (CA), USA, March 2010, paper OWQ6.

[23] N. J. Gomes, M. Morant, A. Alphones, B. Cabon, J. E. Mitchell, C. Lethien, M. Csörnyei, A. Stöhr, and S. Iezekiel, "Radio-over-fiber transport for the support of wireless broadband services Invited," *J. Opt. Netw.,* vol. 8, no. 2, pp. 156-178, 2009.

[24] J. E. Mitchell, "Radio over fibre networks: Advances and challenges," *European Conference and Exhibition on Optical Communication (ECOC)*, Vienna, Austria, September 2009.

[25] A. Ng'oma, "Radio-over-fibre Technology for Broadband Wireless Communication Systems," PhD Thesis, Technische Universiteit Eindhoven, 2005.

[26] C. Cox, III, E. Ackerman, R. Helkey, and G. E. Betts, "Techniques and Performance of Intensity-Modulation Direct-detection analog optical links," *IEEE Trans. Microwave Theory Tech.,* vol. 45, no. 8, pp. 1375-1383, 1997.

[27] R. Irmer, H. Droste, P. Marsch, M. Grieger, G. Fettweis, S. Brueck, H. P. Mayer, L. Thiele, and V. Jungnickel, "Coordinated multipoint: Concepts, performance, and field trial results," *IEEE Commun. Mag.,* vol. 49, no. 2, pp. 102-111, 2011.

[28] W. Kellerer, W. Kiess, L. Scalia, T. Biermann, C. Choi, and K. Kozu, "Novel cellular optical access network and convergence with FTTH," *Optical Fiber Communication Conference (OFC)*, Los Angeles (CA), USA, March 2012, paper OTu2H.3.

[29] L. Juho, K. Younsun, L. Hyojin, N. Boon Loong, D. Mazzarese, L. Jianghua, X. Weimin, and Z. Yongxing, "Coordinated multipoint transmission and reception in LTE-advanced systems," *IEEE Commun. Mag.,* vol. 50, no. 11, pp. 44-50, 2012.

[30] T. Pfeiffer, "Converged heterogeneous optical metro-access networks," *European Conference and Exhibition on Optical Communication (ECOC)*, Torino, Italy, 2010.

[31] T. Pfeiffer, "New avenues of revenues - Open access and infrastructure virtualization," *Optical Fiber Communication Conference (OFC)*, Los Angeles (CA), USA, March 2012, paper NTh4E.1.

[32] L. H. Spiekman, J. M. Wiesenfeld, A. H. Gnauck, L. D. Garrett, G. N. Van den Hoven, T. Van Dongen, M. J. H. Sander-Jochem, and J.

J. M. Binsma, "8 x 10 Gb/s DWDM transmission over 240 km of standard fiber using a cascade of semiconductor optical amplifiers," *IEEE Photon. Technol. Lett.* , vol. 12, no. 8, pp. 1082-1084, 2000.

[33] N. Antoniades, K. C. Reichmann, P. P. Iannone, and A. M. Levine, "Engineering methodology for the use of SOAs and CWDM transmission in the metro network environment," *Optical Fiber Communication Conference (OFC)*, Anaheim (CA), USA, March 2006, paper OTuG6.

[34] S. Liu, K. A. Williams, T. Lin, M. G. Thompson, C. K. Yow, A. Wonfor, R. V. Penty, I. H. White, F. Hopfer, M. Lammlin, and D. Bimberg, "Cascaded performance of quantum dot semiconductor optical amplifier in a recirculating loop," *Conference on Lasers and Electro-Optics (CLEO) and on Quantum Electronics and Laser Science (QELS)*, Long Beach (CA), USA, May 2006, paper CTuM4.

[35] P. J. Winzer, and R. Essiambre, "Advanced Optical Modulation Formats," *Proc. IEEE,* vol. 94, no. 5, pp. 952-985, 2006.

[36] ITU, "Recommendation P.676-9, Attenuation by atmospheric gases," 2012.

[37] ITU, "Recommendation P.840-4, Attenuation due to clouds and fog," 2009.

[38] ITU, "Recommendation P.838-3, Specific Attenuation Model for Rain for Use in Prediction Methods," 2005.

[39] K. Su, L. Moeller, R. B. Barat, and J. F. Federici, "Experimental comparison of performance degradation from terahertz and infrared wireless links in fog," *J. Opt. Soc. Am. A,* vol. 29, no. 2, pp. 179-184, 2012.

[40] I. I. Kim, B. McArthur, and E. J. Korevaar, "Comparison of laser beam propagation at 785 nm and 1550 nm in fog and haze for optical wireless communications," *Proc. SPIE 4214, Optical Wireless Communications III*, Boston (MA), USA, February 2001, pp. 26-37.

[41] I. Kallfass, J. Antes, T. Schneider, F. Kurz, D. Lopez-Diaz, S. Diebold, H. Massler, A. Leuther, and A. Tessmann, "All Active MMIC-Based Wireless Communication at 220 GHz," *IEEE Trans. THz Sci. Technol.,* vol. 1, no. 2, pp. 477-487, 2011.

[42] H. Ito, S. Kodama, Y. Muramoto, T. Furuta, T. Nagatsuma, and T. Ishibashi, "High-speed and high-output InP-InGaAs unitraveling-carrier photodiodes," *IEEE J. Sel. Topics Quantum Electron.,* vol. 10, no. 4, pp. 709-727, 2004.

[43] A. Stöhr, "10 Gbit/s wireless transmission using millimeter-wave over optical fiber systems," *Optical Fiber Communication Conference (OFC)*, Los Angeles (CA), USA, March 2011, paper OTuO3.

[44] A. Stöhr, A. Akrout, R. Buß, B. Charbonnier, F. van Dijk, A. Enard, S. Fedderwitz, D. Jäger, M. Huchard, F. Lecoche, J. Marti, R. Sambaraju, A. Steffan, A. Umbach, and M. Weiß, "60 GHz radio-over-fiber technologies for broadband wireless services Invited," *J. Opt. Netw.,* vol. 8, no. 5, pp. 471-487, 2009.

[45] A. Stöhr, S. Babiel, P. J. Cannard, B. Charbonnier, F. van Dijk, S. Fedderwitz, D. Moodie, L. Pavlovic, L. Ponnampalam, C. C. Renaud, D. Rogers, V. Rymanov, A. Seeds, A. G. Steffan, A. Umbach, and M. Weiss, "Millimeter-Wave Photonic Components for Broadband Wireless Systems," *IEEE Trans. Microwave Theory Tech.,* vol. 58, no. 11, pp. 3071-3082, 2010.

[46] M. Weiss, A. Stöhr, F. Lecoche, and B. Charbonnier, "27 Gbit/s photonic wireless 60 GHz transmission system using 16-QAM OFDM," *International Topical Meeting on Microwave Photonics(MWP)*, Valencia, Spain, 2009.

[47] M. Weiss, M. Huchard, A. Stöhr, B. Charbonnier, S. Fedderwitz, and D. S. Jäger, "60-GHz Photonic Millimeter-Wave Link for Short- to Medium-Range Wireless Transmission Up to 12.5 Gb/s," *J. Lightw. Technol.,* vol. 26, no. 15, pp. 2424-2429, 2008.

[48] C.-T. Lin, A. Ng'oma, W.-Y. Lee, C.-C. Wei, C.-Y. Wang, T.-H. Lu, J. Chen, W. Jiang, Jr., and C.-H. Ho, "2 x 2 MIMO radio-over-fiber system at 60 GHz employing frequency domain equalization," *Opt. Express,* vol. 20, no. 1, pp. 562-567, 2012.

[49] W. Jiang, Jr., C.-T. Lin, L.-Y. W. He, C.-C. Wei, C.-H. Ho, Y.-M. Yang, P.-T. Shih, J. Chen, and S. Chi, "32.65-Gbps OFDM RoF signal generation at 60GHz employing an adaptive I/Q imbalance correction," *European Conference and Exhibition on Optical Communication (ECOC)*, Torino, Italy, 2010.

[50] C.-H. Ho, R. Sambaraju, W. Jiang, Jr., T. H. Lu, C.-Y. Wang, H. Yang, W.-Y. Lee, C.-T. Lin, C.-C. Wei, S. Chi, and A. Ng'oma, "50-Gb/s Radio-over-Fiber System Employing MIMO and OFDM Modulation at 60 GHz," *Optical Fiber Communication Conference (OFC)*, Los Angeles (CA), USA, March 2012, paper OM2B.3.

[51] W. Jiang, Jr., H. Yang, Y.-M. Yang, C.-T. Lin, and A. Ng'oma, "40 Gb/s RoF signal transmission with 10 m wireless distance at 60 GHz," *Optical Fiber Communication Conference (OFC)*, Los Angeles (CA), USA, March 2012, paper OTu2H.1.

[52] A. Ng'oma, C.-T. Lin, L.-Y. W. He, W. Jiang, Jr., F. Annunziata, J. Chen, P.-T. Shih, J. George, and S. Chi, "31 Gbps RoF system employing adaptive bit-loading OFDM modulation at 60 GHz," *Optical Fiber Communication Conference (OFC)*, Los Angeles (CA), USA, March 2011, paper OWT7.

[53] A. Ng'oma, P.-T. Shih, J. George, F. Annunziata, M. Sauer, C.-T. Lin, W. Jiang, Jr., J. Chen, and S. Chi, "21 Gbps OFDM wireless signal transmission at 60 GHz using a simple IMDD radio-over-fiber system," *Optical Fiber Communication Conference (OFC)*, San Diego (CA), USA, March 2010, paper OTuF4.

[54] R. W. Ridgway, and D. W. Nippa, "Generation and Modulation of a 94-GHz Signal Using Electrooptic Modulators," *IEEE Photon. Technol. Lett.,* vol. 20, no. 8, pp. 653-655, 2008.

[55] C. W. Chow, C. H. Yeh, C. B. Huang, J. W. Shi, and C. L. Pan, "Optical carrier distributed network with 0.1 THz short-reach wireless communication system," *Optical Fiber Communication Conference (OFC)*, Los Angeles (CA), USA, March 2012, paper OTu2H.2.

[56] N.-W. Chen, J.-W. Shi, H.-J. Tsai, J.-M. Wun, F.-M. Kuo, J. Hesler, T. W. Crowe, and J. E. Bowers, "Design and demonstration of ultra-fast W-band photonic transmitter-mixer and detectors for 25 Gbits/sec error-free wireless linking," *Opt. Express,* vol. 20, no. 19, pp. 21223-21234, 2012.

[57] N.-W. Chen, H.-J. Tsai, F.-M. Kuo, and J.-W. Shi, "High-Speed W - Band Integrated Photonic Transmitter for Radio-Over-Fiber Applications," *IEEE Trans. Microwave Theory Tech.,* vol. 59, no. 4, pp. 978-986, 2011.

[58] F.-M. Kuo, N.-W. Chen, H.-J. Tsai, J.-W. Shi, and J. Bowers, "High-Power Photonic Transmitter-Mixers with Integrated Wilkinson Power Combiner for Wireless Communication with High Data Rate (15 Gbps) at W-band," *Optical Fiber Communication Conference (OFC)*, Los Angeles (CA), USA, 2012, paper OTh1E.7.

[59] F. M. Kuo, J. W. Shi, N.-W. Chen, C. B. Huang, H. P. Chuang, H.-J. Tsai, and C.-L. Pan, "20-Gb/s error-free wireless transmission using ultra-wideband photonic transmitter-mixer excited with remote distributed optical pulse train," *Optical Fiber Communication Conference (OFC)*, Los Angeles (CA), USA, March 2011, paper OWT5.

[60] F. M. Kuo, C. B. Huang, J. W. Shi, N.-W. Chen, H. P. Chuang, J. E. Bowers, and C.-L. Pan, "Remotely Up-Converted 20-Gbit/s Error-Free Wireless On-Off-Keying Data Transmission at W-Band Using an Ultra-Wideband Photonic Transmitter-Mixer," *IEEE Photon. J.,* vol. 3, no. 2, pp. 209-219, 2011.

[61] A. Kanno, K. Inagaki, I. Morohashi, T. Sakamoto, T. Kuri, I. Hosako, T. Kawanishi, Y. Yoshida, and K. Kitayama, "40 Gb/s W-band (75-110 GHz) 16-QAM radio-over-fiber signal generation and its wireless transmission," *Opt. Express,* vol. 19, no. 26, pp. B56-B63, 2011.

[62] M. Beltran, L. Deng, X. Pang, X. Zhang, V. Arlunno, Y. Zhao, X. Yu, R. Llorente, D. Liu, and I. T. Monroy, "38.2-Gb/s Optical-Wireless Transmission in 75-110 GHz Based on Electrical OFDM with Optical Comb Expansion," *Optical Fiber Communication Conference (OFC)*, Los Angeles (CA), USA, March 2012, paper OM2B.2.

[63] L. Deng, D. Liu, X. Pang, X. Zhang, V. Arlunno, Y. Zhao, A. Caballero, A. K. Dogadaev, X. Yu, I. T. Monroy, M. Beltran, and R. Llorente, "42.13 gbit/s 16qam-OFDM photonics-wireless transmission in 75-110 GHz band," *Progress In Electromagnetics Research,* vol. 126, pp. 449-461, 2012.

[64] D. Zibar, R. Sambaraju, A. Caballero, J. Herrera, and I. T. Monroy, "Carrier Recovery and Equalization for Photonic-Wireless Links with Capacities up to 40 Gb/s in 75-110 GHz Band," *Optical Fiber Communication Conference (OFC)*, Los Angeles (CA), USA, March 2011, paper OThJ4.

[65] R. Sambaraju, D. Zibar, A. Caballero, J. Herrera, J. B. Jensen, I. T. Monroy, U. Westergen, A. Walber, and J. Marti, "Up to 40Gb/s wireless signal generation and demodulation in 75–110 GHz band," *International Topical Meeting on Microwave Photonics (MWP)*, Valencia, Spain, 2010, postdeadline paper.

[66] R. Sambaraju, D. Zibar, A. Caballero, I. T. Monroy, R. Alemany, and J. Herrera, "100-GHz Wireless-Over-Fiber Links With Up to 16-Gb/s QPSK Modulation Using Optical Heterodyne Generation and Digital Coherent Detection," *IEEE Photon. Technol. Lett.,* vol. 22, no. 22, pp. 1650-1652, 2010.

[67] A. Kanno, T. Kuri, I. Hosako, T. Kawanishi, Y. Yoshida, Y. Yasumura, and K. Kitayama, "Optical and millimeter-wave radio seamless MIMO transmission based on a radio over fiber technology," *Opt. Express,* vol. 20, no. 28, pp. 29395-29403, 2012.

[68] X. Pang, A. Caballero, A. Dogadaev, V. Arlunno, R. Borkowski, J. S. Pedersen, L. Deng, F. Karinou, F. Roubeau, D. Zibar, X. Yu, and I. T. Monroy, "100 Gbit/s hybrid optical fiber-wireless link in the W-band (75-110 GHz)," *Opt. Express,* vol. 19, no. 25, pp. 24944-24949, 2011.

[69] X. Li, Z. Dong, J. Yu, N. Chi, Y. Shao, and G. K. Chang, "Fiber-wireless transmission system of 108 Gb/s data over 80 km fiber and 2×2 multiple-input multiple-output wireless links at 100 GHz W-band frequency," *Opt. Lett.,* vol. 37, no. 24, pp. 5106-5108, 2012.

[70] H. J. Song, K. Ajito, A. Hirata, A. Wakatsuki, Y. Muramoto, T. Furuta, N. Kukutsu, T. Nagatsuma, and Y. Kado, "8 Gbit/s wireless data transmission at 250 GHz," *Electron. Lett.,* vol. 45, no. 22, pp. 1121-1122, 2009.

[71] G. Ducournau, P. Szriftgiser, D. Bacquet, A. Beck, T. Akalin, E. Peytavit, M. Zaknoune, and J. F. Lampin, "Optically power supplied Gbit/s wireless hotspot using 1.55μm THz photomixer and heterodyne detection at 200 GHz," *Electron. Lett.,* vol. 46, no. 19, pp. 1349-1351, 2010.

[72] A. Hirata, T. Kosugi, H. Takahashi, R. Yamaguchi, F. Nakajima, T. Furuta, H. Ito, H. Sugahara, Y. Sato, and T. Nagatsuma, "120-GHz-band millimeter-wave photonic wireless link for 10-Gb/s data

transmission," *IEEE Trans. Microwave Theory Tech.,* vol. 54, no. 5, pp. 1937-1944, 2006.

[73] H. J. Song, K. Ajito, Y. Muramoto, A. Wakatsuki, T. Nagatsuma, and N. Kukutsu, "24 Gbit/s data transmission in 300 GHz band for future terahertz communications," *Electron. Lett.,* vol. 48, no. 15, pp. 953-954, 2012.

[74] A. Hirata, T. Kosugi, H. Takahashi, J. Takeuchi, K. Murata, N. Kukutsu, Y. Kado, S. Okabe, T. Ikeda, F. Suginosita, K. Shogen, H. Nishikawa, A. Irino, T. Nakayama, and N. Sudo, "5.8-km 10-Gbps data transmission over a 120-GHz-band wireless link," *IEEE International Conference on Wireless Information Technology and Systems (ICWITS)*, Honolulu (HI), USA, August 2010.

[75] C. E. Shannon, "A mathematical theory of communication," *The Bell System Technical Journal,* vol. 27, no. 3, pp. 379-423, 1948.

[76] S. Benedetto, and P. Poggiolini, "Theory of polarization shift keying modulation," *IEEE Trans. Commun.,* vol. 40, no. 4, pp. 708-721, 1992.

[77] S. Benedetto, and P. T. Poggiolini, "Multilevel polarization shift keying: optimum receiver structure and performance evaluation," *IEEE Trans. Commun.,* vol. 42, no. 234, pp. 1174-1186, 1994.

[78] S. Benedetto, R. Gaudino, and P. Poggiolini, "Polarization recovery in optical polarization shift-keying systems," *IEEE Trans. Commun.,* vol. 45, no. 10, pp. 1269-1279, 1997.

[79] P. M. Hill, R. Olshansky, and W. K. Burns, "Optical polarization division multiplexing at 4 Gb/s," *IEEE Photon. Technol. Lett.,* vol. 4, no. 5, pp. 500-502, 1992.

[80] S. Tsukamoto, D. Ly-Gagnon, K. Katoh, and K. Kikuchi, "Coherent demodulation of 40-Gbit/s polarization-multiplexed QPSK signals with 16-GHz spacing after 200-km transmission," *Optical Fiber Communication Conference (OFC)*, Anaheim (CA), USA, March 2005, postdeadline paper PDP29.

[81] K.-D. Kammeyer, *Nachrichtenübertragung*, 4th ed., Vieweg + Teubner, 2008.

[82] J. Proakis, *Digital Communications*, 3rd ed., McGraw-Hill, 1995.

[83] R. Schmogrow, D. Hillerkuss, M. Dreschmann, M. Huebner, M. Winter, J. Meyer, B. Nebendahl, C. Koos, J. Becker, W. Freude, and J. Leuthold, "Real-Time Software-Defined Multiformat Transmitter Generating 64QAM at 28 GBd," *IEEE Photon. Technol. Lett.,* vol. 22, no. 21, pp. 1601-1603, 2010.

[84] J. Chuang, and N. Sollenberger, "Beyond 3G: wideband wireless data access based on OFDM and dynamic packet assignment," *IEEE Commun. Mag.,* vol. 38, no. 7, pp. 78-87, 2000.

[85] I. Koffman, and V. Roman, "Broadband wireless access solutions based on OFDM access in IEEE 802.16," *IEEE Commun. Mag.,* vol. 40, no. 4, pp. 96-103, 2002.

[86] W. Zhendao, and G. B. Giannakis, "Wireless multicarrier communications," *IEEE Signal Process. Mag.,* vol. 17, no. 3, pp. 29-48, 2000.

[87] J. Armstrong, "OFDM for Optical Communications," *J. Lightw. Technol.,* vol. 27, no. 3, pp. 189-204, 2009.

[88] R. Schmogrow, M. Winter, D. Hillerkuss, B. Nebendahl, S. Ben-Ezra, J. Meyer, M. Dreschmann, M. Huebner, J. Becker, C. Koos, W. Freude, and J. Leuthold, "Real-time OFDM transmitter beyond 100 Gbit/s," *Opt. Express,* vol. 19, no. 13, pp. 12740-12749, 2011.

[89] R. Schmogrow, R. Bouziane, M. Meyer, P. A. Milder, P. C. Schindler, R. I. Killey, P. Bayvel, C. Koos, W. Freude, and J. Leuthold, "Real-time OFDM or Nyquist pulse generation - which performs better with limited resources?," *Opt. Express,* vol. 20, no. 26, pp. B543-B551, 2012.

[90] R. Bouziane, R. Schmogrow, D. Hillerkuss, P. A. Milder, C. Koos, W. Freude, J. Leuthold, P. Bayvel, and R. I. Killey, "Generation and transmission of 85.4 Gb/s real-time 16QAM coherent optical OFDM signals over 400 km SSMF with preamble-less reception," *Opt. Express,* vol. 20, no. 19, pp. 21612-21617, 2012.

[91] A. J. Lowery, and L. B. Du, "Optical orthogonal division multiplexing for long haul optical communications: A review of the first five years," *Optical Fiber Technology,* vol. 17, no. 5, pp. 421-438, 2011.

[92] H. Ochiai, and H. Imai, "On the distribution of the peak-to-average power ratio in OFDM signals," *IEEE Trans. Commun.,* vol. 49, no. 2, pp. 282-289, 2001.

[93] H. Seung Hee, and L. Jae Hong, "An overview of peak-to-average power ratio reduction techniques for multicarrier transmission," *IEEE Wireless Communications,* vol. 12, no. 2, pp. 56-65, 2005.

[94] J. E. Bowers, and C. A. Burrus, "High-speed zero-bias waveguide photodetectors," *Electron. Lett.,* vol. 22, no. 17, pp. 905-906, 1986.

[95] K. Kato, "Ultrawide-band/high-frequency photodetectors," *IEEE Trans. Microwave Theory Tech.,* vol. 47, no. 7, pp. 1265-1281, 1999.

[96] R. Noe, "Phase noise-tolerant synchronous QPSK/BPSK baseband-type intradyne receiver concept with feedforward carrier recovery," *J. Lightw. Technol.,* vol. 23, no. 2, pp. 802-808, 2005.

[97] A. Leven, N. Kaneda, K. Ut-Va, and C. Young-Kai, "Frequency Estimation in Intradyne Reception," *IEEE Photon. Technol. Lett. ,* vol. 19, no. 6, pp. 366-368, 2007.

[98] T. Pfau, S. Hoffmann, R. Peveling, S. Bhandare, S. K. Ibrahim, O. Adamczyk, M. Porrmann, R. Noe, and Y. Achiam, "First Real-Time Data Recovery for Synchronous QPSK Transmission With Standard DFB Lasers," *IEEE Photon. Technol. Lett. ,* vol. 18, no. 18, pp. 1907-1909, 2006.

[99] M. J. Connelly, *Semiconductor Optical Amplifiers*, 1st ed., New York, Boston, Dordrecht, London, Moscow: Kluwer Academic Publishers, 2002.

[100] N. K. Dutta, and Q. Wang, *Semiconductor optical amplifiers*, 1st ed.: World Scientific Publishing Co Pte Ltd, 2006.

[101] N. A. Olsson, "Lightwave systems with optical amplifiers," *J. Lightw. Technol.,* vol. 7, no. 7, pp. 1071-1082, 1989.

[102] C. Meuer, J. Kim, M. Laemmlin, S. Liebich, A. Capua, G. Eisenstein, A. R. Kovsh, S. S. Mikhrin, I. L. Krestnikov, and D. Bimberg, "Static gain saturation in quantum dot semiconductor optical amplifiers," *Opt. Express,* vol. 16, no. 11, pp. 8269-8279, 2008.

[103] G.-W. Lu, M. Sköld, P. Johannisson, J. Zhao, M. Sjödin, H. Sunnerud, M. Westlund, A. Ellis, and P. A. Andrekson, "40-Gbaud 16-QAM

transmitter using tandem IQ modulators with binary driving electronic signals," *Opt. Express,* vol. 18, no. 22, pp. 23062-23069, 2010.

[104] T. Vallaitis, C. Koos, R. Bonk, W. Freude, M. Laemmlin, C. Meuer, D. Bimberg, and J. Leuthold, "Slow and fast dynamics of gain and phase in a quantum dot semiconductor optical amplifier," *Opt. Express,* vol. 16, no. 1, pp. 170-178, 2008.

[105] W. Jin, A. Maitra, C. G. Poulton, W. Freude, and J. Leuthold, "Temporal Dynamics of the Alpha Factor in Semiconductor Optical Amplifiers," *J. Lightw. Technol.,* vol. 25, no. 3, pp. 891-900, 2007.

[106] A. J. Seeds, and K. J. Williams, "Microwave Photonics," *J. Lightw. Technol.,* vol. 24, no. 12, pp. 4628-4641, 2006.

[107] J. Capmany, and D. Novak, "Microwave photonics combines two worlds," *Nature Photon.,* vol. 1, no. 6, pp. 319-330, 2007.

[108] S. Iezekiel, *Microwave Photonics - Devices and Applications*, John Wiley & Sons, Ltd, 2009.

[109] C. H. Cox, III, E. I. Ackerman, G. E. Betts, and J. L. Prince, "Limits on the performance of RF-over-fiber links and their impact on device design," *IEEE Trans. Microwave Theory Tech.,* vol. 54, no. 2, pp. 906-920, 2006.

[110] U. Gliese, S. Norskov, and T. N. Nielsen, "Chromatic dispersion in fiber-optic microwave and millimeter-wave links," *IEEE Trans. Microwave Theory Tech.,* vol. 44, no. 10, pp. 1716-1724, 1996.

[111] C. Lim, A. Nirmalathas, M. Bakaul, K.-L. Lee, D. Novak, and R. Waterhouse, "Mitigation strategy for transmission impairments in millimeter-wave radio-over-fiber networks Invited," *J. Opt. Netw.,* vol. 8, no. 2, pp. 201-214, 2009.

[112] G. H. Smith, D. Novak, and Z. Ahmed, "Technique for optical SSB generation to overcome dispersion penalties in fibre-radio systems," *Electron. Lett.,* vol. 33, no. 1, pp. 74-75, 1997.

[113] G. H. Smith, D. Novak, and Z. Ahmed, "Overcoming chromatic-dispersion effects in fiber-wireless systems incorporating external modulators," *IEEE Trans. Microwave Theory Tech.,* vol. 45, no. 8, pp. 1410-1415, 1997.

[114] D. Wake, S. Dupont, C. Lethien, J. P. Vilcot, and D. Decoster, "Radiofrequency transmission of 32-QAM signals over multimode fibre for distributed antenna system applications," *Electron. Lett.,* vol. 37, no. 17, pp. 1087-1089, 2001.

[115] M. Sauer, A. Kobyakov, and J. George, "Radio Over Fiber for Picocellular Network Architectures," *J. Lightw. Technol.,* vol. 25, no. 11, pp. 3301-3320, 2007.

[116] N. J. Gomes, A. Nkansah, and D. Wake, "Radio-Over-MMF Techniques—Part I: RF to Microwave Frequency Systems," *J. Lightw. Technol.,* vol. 26, no. 15, pp. 2388-2395, 2008.

[117] A. M. J. Koonen, and L. M. Garcia, "Radio-Over-MMF Techniques—Part II: Microwave to Millimeter-Wave Systems," *J. Lightw. Technol.,* vol. 26, no. 15, pp. 2396-2408, 2008.

[118] J. James, P. Shen, A. Nkansah, X. Liang, and N. J. Gomes, "Millimeter-Wave Wireless Local Area Network Over Multimode Fiber System Demonstration," *IEEE Photon. Technol. Lett.,* vol. 22, no. 9, pp. 601-603, 2010.

[119] R. P. Braun, G. Grosskopf, D. Rohde, and F. Schmidt, "Optical millimetre-wave generation and transmission experiments for mobile 60 GHz band communications," *Electron. Lett.,* vol. 32, no. 7, pp. 626-628, 1996.

[120] L. Noel, D. Wake, D. G. Moodie, D. D. Marcenac, L. D. Westbrook, and D. Nesset, "Novel techniques for high-capacity 60-GHz fiber-radio transmission systems," *IEEE Trans. Microwave Theory Tech.,* vol. 45, no. 8, pp. 1416-1423, 1997.

[121] K. Kitayama, "An approach to single optical component antenna base stations for broad-band millimeter-wave fiber-radio access systems," *IEEE Trans. Microwave Theory Tech.,* vol. 48, pp. 2588-2595, 2000.

[122] T. Kuri, K. Kitayama, and Y. Takahashi, "60-GHz-band full-duplex radio-on-fiber system using two-RF-port electroabsorption transceiver," *IEEE Photon. Technol. Lett.,* vol. 12, no. 4, pp. 419-421, 2000.

[123] J. J. Vegas Olmos, T. Kuri, T. Sono, K. Tamura, H. Toda, and K. i. Kitayama, "Reconfigurable 2.5-Gb/s Baseband and 60-GHz

(155-Mb/s) Millimeter-Waveband Radio-Over-Fiber (Interleaving) Access Network," *J. Lightw. Technol.,* vol. 26, no. 15, pp. 2506-2512, 2008.

[124] N. Pleros, K. Vyrsokinos, K. Tsagkaris, and N. D. Tselikas, "A 60 GHz Radio-Over-Fiber Network Architecture for Seamless Communication With High Mobility," *J. Lightw. Technol.,* vol. 27, no. 12, pp. 1957-1967, 2009.

[125] R. Svitek, and S. Raman, "DC offsets in direct-conversion receivers: characterization and implications," *IEEE Microw. Mag.,* vol. 6, no. 3, pp. 76-86, 2005.

[126] M. Valkama, M. Renfors, and V. Koivunen, "Advanced methods for I/Q imbalance compensation in communication receivers," *IEEE Trans. Signal Process.,* vol. 49, no. 10, pp. 2335-2344, 2001.

[127] M. Windisch, and G. Fettweis, "Blind I/Q imbalance parameter estimation and compensation in low-IF receivers," *International Symposium on Control, Communications and Signal Processing,* Hammamet, Tunisia, March 2004, pp. 75-78.

[128] G. J. Schneider, J. A. Murakowski, C. A. Schuetz, S. Shi, and D. W. Prather, "Radiofrequency signal-generation system with over seven octaves of continuous tuning," *Nature Photon.,* vol. 7, no. 2, pp. 118-122, 2013.

[129] N. Yasuoka, H. Ebe, K. Kawaguchi, M. Ekawa, S. Sekiguchi, K. Morito, O. Wada, M. Sugawara, and Y. Arakawa, "Polarization-Insensitive Quantum Dot Semiconductor Optical Amplifiers Using Strain-Controlled Columnar Quantum Dots," *J. Lightw. Technol.,* vol. 30, no. 1, pp. 68-75, 2012.

[130] D. R. Zimmerman, and L. H. Spiekman, "Amplifiers for the masses: EDFA, EDWA, and SOA amplets for metro and access applications," *J. Lightw. Technol.,* vol. 22, no. 1, pp. 63-70, 2004.

[131] H. Schmuck, R. Bonk, W. Poehlmann, C. Haslach, W. Kuebart, D. Karnick, J. Meyer, D. Fritzsche, E. Weis, J. Becker, W. Freude, and T. Pfeiffer, "Demonstration of an SOA-assisted open metro-access infrastructure for heterogeneous services," *Opt. Express,* vol. 22, no. 1, pp. 737-748, 2014.

[132] J. Leuthold, R. Ryf, D. N. Maywar, S. Cabot, J. Jaques, and S. S. Patel, "Nonblocking all-optical cross connect based on regenerative all-optical wavelength converter in a transparent demonstration over 42 nodes and 16800 km," *J. Lightw. Technol.,* vol. 21, no. 11, pp. 2863-2870, 2003.

[133] G. Contestabile, "All optical processing in QD-SOAs," *Optical Fiber Communication Conference (OFC)*, San Francisco (CA), USA, March 2014, paper W4F.6.

[134] W. Freude, R. Bonk, T. Vallaitis, A. Marculescu, A. Kapoor, E. K. Sharma, C. Meuer, D. Bimberg, R. Brenot, F. Lelarge, D. Guang-Hua, C. Koos, and J. Leuthold, "Linear and nonlinear semiconductor optical amplifiers," *International Conference on Transparent Optical Networks (ICTON)*, Munich, Germany, July 2010.

[135] J. Leuthold, "Linear and Nonlinear Semiconductor Optical Amplifiers," *Optical Fiber Communication Conference (OFC)*, San Diego, CA, USA, March 2010, paper OThI3.

[136] R. Bonk, T. Vallaitis, J. Guetlein, C. Meuer, H. Schmeckebier, D. Bimberg, C. Koos, W. Freude, and J. Leuthold, "The Input Power Dynamic Range of a Semiconductor Optical Amplifier and Its Relevance for Access Network Applications," *IEEE Photon. J.,* vol. 3, no. 6, pp. 1039-1053, 2011.

[137] R. J. Manning, D. A. O. Davies, and J. K. Lucek, "Recovery rates in semiconductor laser amplifiers: optical and electrical bias dependencies," *Electron. Lett.,* vol. 30, no. 15, pp. 1233-1235, 1994.

[138] D. Wolfson, S. L. Danielsen, C. Joergensen, B. Mikkelsen, and K. E. Stubkjaer, "Detailed theoretical investigation of the input power dynamic range for gain-clamped semiconductor optical amplifier gates at 10 Gb/s," *IEEE Photon. Technol. Lett.,* vol. 10, no. 9, pp. 1241-1243, 1998.

[139] D. A. Francis, S. P. DiJaili, and J. D. Walker, "A single-chip linear optical amplifier," *Optical Fiber Communication Conference (OFC)*, Anaheim (CA), USA, March 2001, postdeadline paper PDP13.

[140] C. Michie, A. E. Kelly, I. Armstrong, I. Andonovic, and C. Tombling, "An Adjustable Gain-Clamped Semiconductor Optical Amplifier

(AGC-SOA)," *J. Lightw. Technol.,* vol. 25, no. 6, pp. 1466-1473, 2007.

[141] T. Hung Nguyen, M. Matsuura, and N. Kishi, "Enhancement of Input Power Dynamic Range for Multiwavelength Amplification and Optical Signal Processing in a Semiconductor Optical Amplifier Using Holding Beam Effect," *J. Lightw. Technol.,* vol. 28, no. 17, pp. 2593-2602, 2010.

[142] J. Yu, and P. Jeppesen, "Increasing Input Power Dynamic Range of SOA by Shifting the Transparent Wavelength of Tunable Optical Filter," *J. Lightw. Technol.,* vol. 19, no. 9, pp. 1316, 2001.

[143] J. Leuthold, D. M. Marom, S. Cabot, J. J. Jaques, R. Ryf, and C. R. Giles, "All-Optical Wavelength Conversion Using a Pulse Reformatting Optical Filter," *J. Lightw. Technol.,* vol. 22, no. 1, pp. 186, 2004.

[144] M. Sauer, and J. Hurley, "Experimental 43 Gb/s NRZ and DPSK performance comparison for systems with up to 8 concatenated SOAs," *Conference on Lasers and Electro-Optics (CLEO)*, Long Beach (CA), USA, May 2006, paper CThY2.

[145] E. Ciaramella, A. D'Errico, and V. Donzella, "Using Semiconductor-Optical Amplifiers With Constant Envelope WDM Signals," *IEEE J. Quant. Electron.,* vol. 44, no. 5, pp. 403-409, 2008.

[146] J. D. Downie, and J. Hurley, "Effects of dispersion on SOA nonlinear impairments with DPSK signals," *21st Annual Meeting of the IEEE Lasers and Electro-Optics Society (LEOS)*, Acapulco, Mexico, 2008, pp. 610-611.

[147] P. S. Cho, Y. Achiam, G. Levy-Yurista, M. Margalit, Y. Gross, and J. B. Khurgin, "Investigation of SOA nonlinearities on the amplification of high spectral efficiency signals," *Optical Fiber Communication Conference (OFC)*, Los Angeles (CA), USA, February 2004, paper MF70.

[148] W. Xing, S. Yikai, L. Xiang, J. Leuthold, and S. Chandrasekhar, "10-Gb/s RZ-DPSK transmitter using a saturated SOA as a power booster and limiting amplifier," *IEEE Photon. Technol. Lett.,* vol. 16, no. 6, pp. 1582-1584, 2004.

[149] H. Takeda, N. Hashimoto, T. Akashi, H. Narusawa, K. Matsui, K. Mori, S. Tanaka, and K. Morito, "Wide range over 20 dB output power control using semiconductor optical amplifier for 43.1 Gbps RZ-DQPSK signal," *European Conference and Exhibition on Optical Communication (ECOC)*, Vienna, Austria, 2009.

[150] T. Vallaitis, R. Bonk, J. Guetlein, D. Hillerkuss, J. Li, R. Brenot, F. Lelarge, G. H. Duan, W. Freude, and J. Leuthold, "Quantum dot SOA input power dynamic range improvement for differential-phase encoded signals," *Opt. Express,* vol. 18, no. 6, pp. 6270-6276, 2010.

[151] R. Bonk, G. Huber, T. Vallaitis, R. Schmogrow, D. Hillerkuss, C. Koos, W. Freude, and J. Leuthold, "Impact of alfa-factor on SOA dynamic range for 20 GBd BPSK, QPSK and 16-QAM Signals," *Optical Fiber Communication Conference (OFC)*, los Angeles (CA), USA, March 2011, paper OML4.

[152] N. Kamitani, Y. Yoshida, and K.-i. Kitayama, "Experimental Study on Impact of SOA Nonlinear Phase Noise in 40Gbps Coherent 16QAM Transmissions," *European Conference and Exhibition on Optical Communication (ECOC)*, Amsterdam, Netherlands, September 2012, poster P1.04.

[153] S. Lange, G. Contestabile, Y. Yoshida, and K.-i. Kitayama, "Phase-transparent Amplification of 16 QAM Signals in a QD-SOA," *IEEE Photon. Technol. Lett.,* vol. 25, no. 24, pp. 2486 - 2489, 2013.

[154] S. Lange, Y. Yoshida, and K.-i. Kitayama, "A Low-complexity Digital Pre-compensation of SOA Induced Phase Distortion in Coherent QAM Transmissions," *Optical Fiber Communication Conference (OFC)*, Anaheim (CA), USA, March 2013, paper OTh3C.7.

[155] S. Amiralizadeh, A. T. Nguyen, P. Chul Soo, A. Ghazisaeidi, and L.A. Rusch, "Experimental validation of digital filter back-propagation to suppress SOA-induced nonlinearities in 16-QAM," *Optical Fiber Communication Conference (OFC)*, Anaheim (CA), USA, March 2013, paper OM2B.2.

[156] S. Liu, K. A. Williams, T. Lin, M. G. Thompson, C. K. Yow, A. Wonfor, R. V. Penty, I. H. White, F. Hopfer, M. Lammlin, and D. Bimberg, "Cascaded performance of quantum dot semiconductor optical amplifier in a recirculating loop," *Conference on Lasers and*

Electro-Optics (CLEO) and on Quantum Electronics and Laser Science (QELS), Long Beach (CA), USA, May 2006, paper CTuM4.

[157] A. R. Kovsh, N. A. Maleev, A. E. Zhukov, S. S. Mikhrin, A. P. Vasil'ev, E. A. Semenova, Y. M. Shernyakov, M. V. Maximov, D. A. Livshits, V. M. Ustinov, N. N. Ledentsov, D. Bimberg, and Z. I. Alferov, "InAs/InGaAs/GaAs quantum dot lasers of 1.3μm range with enhanced optical gain," *Journal of Crystal Growth,* vol. 251, no. 1–4, pp. 729-736, 2003.

[158] V. J. Urick, J. X. Qiu, and F. Bucholtz, "Wide-band QAM-over-fiber using phase modulation and interferometric demodulation," *IEEE Photon. Technol. Lett.,* vol. 16, no. 10, pp. 2374-2376, 2004.

[159] S. B. Weinstein, "The history of orthogonal frequency-division multiplexing [History of Communications]," *IEEE Commun. Mag.,* vol. 47, no. 11, pp. 26-35, 2009.

[160] S. Cherry, "Edholm's law of bandwidth," *IEEE Spectr.,* vol. 41, no. 7, pp. 58-60, 2004.

[161] E. Ciaramella, Y. Arimoto, G. Contestabile, M. Presi, A. D'Errico, V. Guarino, and M. Matsumoto, "1.28 terabit/s (32x40 Gbit/s) wdm transmission system for free space optical communications," *IEEE J. Sel. Areas Commun.,* vol. 27, no. 9, pp. 1639-1645, 2009.

[162] K. Su, L. Moeller, R. B. Barat, and J. F. Federici, "Experimental comparison of terahertz and infrared data signal attenuation in dust clouds," *J. Opt. Soc. Am. A,* vol. 29, no. 11, pp. 2360-2366, 2012.

[163] H.-C. Chien, A. Chowdhury, Z. Jia, Y.-T. Hsueh, and G.-K. Chang, "60 GHz millimeter-wave gigabit wireless services over long-reach passive optical network using remote signal regeneration and upconversion," *Opt. Express,* vol. 17, no. 5, pp. 3016-3024, 2009.

[164] A. Hirata, H. Takahashi, R. Yamaguchi, T. Kosugi, K. Murata, T. Nagatsuma, N. Kukutsu, and Y. Kado, "Transmission Characteristics of 120-GHz-Band Wireless Link Using Radio-on-Fiber Technologies," *J. Lightw. Technol.,* vol. 26, no. 15, pp. 2338-2344, 2008.

[165] H. Ito, T. Furuta, Y. Muramoto, T. Ito, and T. Ishibashi, "Photonic millimetre- and sub-millimetrewave generation using J-band

rectangularwaveguide-output uni-travelling-carrier photodiode module," *Electron. Lett.,* vol. 42, no. 24, pp. 1424-1425, 2006.

[166] R. Yamaguchi, A. Hirata, T. Kosugi, H. Takahashi, N. Kukutsu, T. Nagatsuma, Y. Kado, H. Ikegawa, H. Nishikawa, and T. Nakayama, "10-Gbit/s MMIC wireless link exceeding 800 meters," *IEEE Radio and Wireless Symposium*, 2008, pp. 695-698.

[167] A. Leuther, A. Tessmann, I. Kallfass, H. Massler, R. Loesch, M. Schlechtweg, M. Mikulla, and O. Ambacher, "Metamorphic HEMT technology for submillimeter-wave MMIC applications," *International Conference on Indium Phosphide & Related Materials (IPRM)*, Kagawa, Japan, May 2010.

[168] I. Kallfass, A. Tessmann, J. Antes, D. Lopez-Diaz, M. Kuri, H. Massler, T. Zwick, and A. Leuther, "Wireless DVB-C transmission at 220 GHz using active single-chip receive and transmit MMICs," *IEEE International Microwave Symposium (IMS)*, Baltimore (MD), USA, June 2011.

[169] D. Lopez-Diaz, I. Kallfass, A. Tessmann, A. Leuther, S. Wagner, M. Schlechtweg, and O. Ambacher, "A subharmonic chipset for gigabit communication around 240 GHz," *IEEE International Microwave Symposium (IMS)*, Montreal, Canada, June 2012.

[170] D. Hillerkuss, T. Schellinger, M. Jordan, C. Weimann, F. Parmigiani, B. Resan, K. Weingarten, S. Ben-Ezra, B. Nebendahl, C. Koos, W. Freude, and J. Leuthold, "High-Quality Optical Frequency Comb by Spectral Slicing of Spectra Broadened by SPM," *IEEE Photon. J.,* vol. 5, no. 5, pp. 7201011-7201011, 2013.

[171] F. Chang, K. Onohara, and T. Mizuochi, "Forward error correction for 100 G transport networks," *IEEE Commun. Mag.,* vol. 48, no. 3, pp. S48-S55, 2010.

[172] T. Kleine-Ostmann, and T. Nagatsuma, "A Review on Terahertz Communications Research," *J. of Infrared, Millimeter and Terahertz Waves,* vol. 32, pp. 143-171, 2011.

[173] M. Tonouchi, "Cutting-edge terahertz technology," *Nature Photon.,* vol. 1, no. 2, pp. 97-105, 2007.

[174] J. Federici, and L. Moeller, "Review of terahertz and subterahertz wireless communications," *Journal of Applied Physics,* vol. 107, no. 11, pp. 111101-22, 2010.

[175] H. J. Song, and T. Nagatsuma, "Present and Future of Terahertz Communications," *IEEE Trans. THz Sci. Technol.,* vol. 1, no. 1, pp. 256-263, 2011.

[176] A. Kanno, K. Inagaki, I. Morohashi, T. Sakamoto, T. Kuri, I. Hosako, T. Kawanishi, Y. Yoshida, and K. ichi Kitayama, "20-Gb/s QPSK W-band (75-110GHz) wireless link in free space using radio-over-fiber technique," *IEICE Electronics Express,* vol. 8, no. 8, pp. 612-617, 2011.

[177] A. Tessmann, A. Leuther, H. Massler, and M. Seelmann-Eggebert, "A High Gain 600 GHz Amplifier TMIC Using 35 nm Metamorphic HEMT Technology," *IEEE Compound Semiconductor Integrated Circuit Symposium (CSICS)*, La Jolla (CA), USA, October 2012.

[178] E. Ip, A. P. T. Lau, D. J. F. Barros, and J. M. Kahn, "Coherent detection in optical fiber systems," *Opt. Express,* vol. 16, no. 2, pp. 753-791, 2008.

[179] D. Ly-Gagnon, S. Tsukamoto, K. Katoh, and K. Kikuchi, "Coherent detection of optical quadrature phase-shift keying signals with carrier phase estimation," *J. Lightw. Technol.,* vol. 24, no. 1, pp. 12-21, 2006.

[180] D. Korn, P. C. Schindler, C. Stamatiadis, M. F. O'Keefe, L. Stampoulidis, R. Schmogrow, P. Zakynthinos, R. Palmer, N. Cameron, Y. Zhou, R. G. Walker, E. Kehayas, I. Tomkos, L. Zimmermann, K. Petermann, W. Freude, C. Koos, and J. Leuthold, "First monolithic GaAs IQ electro-optic modulator, demonstrated at 150 Gbit/s with 64-QAM," *Optical Fiber Communication Conference (OFC)*, Anaheim (CA), USA, March 2013, postdeadline paper PDP5C.4.

[181] G. Fish, "Heterogeneous Photonic Integration for Microwave Photonic Applications," *Optical Fiber Communication Conference (OFC)*, Anaheim (CA), USA, March 2013, paper OW3D.5.

[182] R. Boula-Picard, M. Alouini, J. Lopez, N. Vodjdani, and J.-C. Simon, "Impact of the Gain Saturation Dynamics in Semiconductor Optical

Amplifiers on the Characteristics of an Analog Optical Link," *J. Lightw. Technol.,* vol. 23, no. 8, pp. 2420, 2005.

[183] W. Freude, "Optical Transmitters and Receivers Part 1," lecture notes, Institute of Photonics and Quantum Electronics, Karlsruhe Institute of Technology (KIT), 2012.

[184] C. Jeruchim, "Techniques for Estimating the Bit Error Rate in the Simulation of Digital Communication Systems," *IEEE J. Sel. Areas Commun.,* vol. 2, no. 1, pp. 153-170, 1984.

[185] V. Ribeiro, L. Costa, M. Lima, and A. L. J. Teixeira, "Optical performance monitoring using the novel parametric asynchronous eye diagram," *Opt. Express,* vol. 20, no. 9, pp. 9851-9861, 2012.

[186] A. Papoulis, *Probability, random variables, and stochastic processes*, 3rd ed., New York: McGraw-Hill, 1991.

[187] S. D. Personick, "Receiver Design for Digital Fiber Optic Communication Systems, I," *The Bell System Technical Journal,* vol. 52, no. 6, pp. 843-874, 1973.

[188] S. Okamoto, K. Toyoda, T. Omiya, K. Kasai, M. Yoshida, and M. Nakazawa, "512 QAM (54 Gbit/s) coherent optical transmission over 150 km with an optical bandwidth of 4.1 GHz," *European Conference and Exhibition on Optical Communication (ECOC)*, Torino, Italy, September 2010.

[189] S. Chandrasekhar, and L. Xiang, "Enabling components for future high-speed coherent communication systems," *Optical Fiber Communication Conference (OFC)*, Los Angeles (CA), USA, March 2011, paper OMU5.

[190] Q. Dayou, H. Ming-Fang, E. Ip, H. Yue-Kai, S. Yin, H. Junqiang, and W. Ting, "101.7-Tb/s (370x294-Gb/s) PDM-128QAM-OFDM transmission over 3x55-km SSMF using pilot-based phase noise mitigation," *Optical Fiber Communication Conference (OFC)*, Los Angeles (CA), USA, March 2011, postdeadline paper PDPB5.

[191] R. A. Shafik, S. Rahman, and A. H. M. R. Islam, "On the Extended Relationships Among EVM, BER and SNR as Performance Metrics," *International Conference on Electrical and Computer Engineering (ICECE)*, Dhaka, Bangladesh, December 2006, pp. 408-411.

[192] H. Arslan, and H. A. Mahmoud, "Error vector magnitude to SNR conversion for nondata-aided receivers," *IEEE Trans. Wireless Commun.,* vol. 8, no. 5, pp. 2694-2704, 2009.

[193] M. Müller, R. Stephens, and R. McHugh, "Total jitter measurement at low probability levels, using optimized BERT scan method," White Paper, Agilent Technologies, Santa Clara, 2005.

[194] T. Nagatsuma, H. Ito, and T. Ishibashi, "High-power RF photodiodes and their applications," *Laser & Photonics Reviews,* vol. 3, no. 1-2, pp. 123-137, 2009.

[195] A. Wakatsuki, T. Furuta, Y. Muramoto, and T. Ishibashia, "High-speed photodiode and optical receiver technologies," *Optical Fiber Communication Conference (OFC)*, San Diego (CA), USA, March 2009, paper OMK1.

[196] T. Ishibashi, S. Kodama, N. Shimizu, and T. Furuta, "High-Speed Response of Uni-Traveling-Carrier Photodiodes," *Jpn. J. Appl. Phys.,* vol. 36, No. 10, pp. 6263, 1997.

[197] T. Ishibashi, T. Furuta, H. Fushimi, and H. Ito, "Photoresponse characteristics of uni-traveling-carrier photodiodes," *Proc. SPIE 4283, Physics and Simulation of Optoelectronic Devices IX*, pp. 469-479, 2001.

[198] A. Hirata, H. Togo, N. Shimizu, H. Takahashi, K. Okamoto, and T. Nagatsuma, "Low-Phase Noise Photonic Millimeter-Wave Generator Using an AWG Integrated with a 3-dB Combiner," *IEICE T. Electron.,* vol. E88-C, no. 7, pp. 1458-1464, 2005.

Acknowledgments (German)

Die vorliegende Dissertation entstand während meiner Tätigkeit am Institut für Photonik und Quantenelektronik (IPQ) des Karlsruher Instituts für Technologie (KIT). Sie war zum Teil in die vom Bundesministerium für Bildung und Forschung geförderten BMBF-Projekte MILLILINK und CONDOR eingebunden und wurde außerdem von der Karlsruhe School of Optics and Photonics (KSOP) sowie der Helmholtz International Research School for Teratronics (HIRST) unterstützt.

Am Ende meiner Arbeit möchte ich all denjenigen Personen meinen herzlichen Dank aussprechen, die zum Gelingen des vorliegenden Manuskriptes beigetragen haben.

Mein erster und ganz besonderer Dank gilt meinem Doktorvater Prof. Dr. sc. nat. Jürg Leuthold für das in mich gesetzte Vertrauen und für sein Interesse an meiner Arbeit. Neben den fachlichen Diskussionen mit ihm waren es vor allem seine mutigen Ideen und sein vorausschauendes Denken, welche diese Arbeit mitbeeinflusst und zum Erfolg geführt haben. Besonders die von ihm oftmals wiederholten Schlagworte „See the big picture!“ und „top-down approach“ haben sich in mein Gedächtnis eingeprägt.

Herrn Prof. Dr. Dr. h.c. Wolfgang Freude bin ich zu mindestens ebenso großem Dank verpflichtet. Ich danke ihm für sein kontinuierliches Interesse an meiner Arbeit und für das Lesen und Korrigieren meiner Texte. Sein beständiges und unvergleichliches Streben nach Perfektion, sein unschätzbares Wissen und seine ungebremste Neugierde haben mich stets beeindruckt und dazu beigetragen, hohe Ansprüche an mich selbst zu stellen.

Prof. Dr.-Ing. Ingmar Kallfass danke ich für die freundliche Übernahme des Korreferats und für die außerordentlich erfolgreiche Zusammenarbeit im Projekt MILLILINK unter seiner Leitung. Ich bedanke mich für das mir entgegengebrachte Vertrauen bezüglich der Verwendung der unbezahlbaren MMIC-Module in den zahlreichen Messkampagnen.

Meinem Kollegen Dr. René Bonk danke ich für alles, was ich von ihm lernen durfte, und für alles, was wir in den vielen, vielen Diskussionen erdacht, nochmals überdacht und schließlich in gemeinsamen Projekten und Messungen im Labor erreicht haben. Ich bedanke mich für das blinde Vertrauen und den außerordentlich fairen und sportlichen Umgang miteinander. Ein Dankeschön auch für die Durchsicht des vorliegenden Manuskripts.

Meinem Kollegen Dr. Moritz Röger danke ich für die zahlreichen Diskussionen und Fachsimpeleien über allgemeine elektrotechnische und physikalische Zusammenhänge. Ich danke ihm außerdem für die gemeinsame, kurzweilige Vorbereitung der HLB-Übungen und für das gemeinsame Ersinnen neuer, kniffliger Klausuraufgaben.

Die 100 Gbit/s Funkübertragung in dieser Arbeit wäre ohne die UTC-PD von NTT-NEL nicht möglich gewesen. Ich bedanke mich recht herzlich bei Dr. Shimizu und Dr. Ishibashi für die großzügige Leihgabe und die Unterstützung.

Für die gemeinsamen Messungen und die gute Zusammenarbeit im Projekt MILLILINK danke ich vielmals Jochen Antes (IHE) und Daniel Lopez-Diaz (Fraunhofer IAF Freiburg), deren Sinn für entspanntes Arbeiten mich stets beeindruckt hat. Von beiden habe ich vieles über MMICs und HF-Technik gelernt. Jochen Antes danke ich für die Ko-Betreuung unseres gemeinsamen Masterstudenten Florian Boes. Daniel Lopez-Diaz danke ich für den Entwurf des großartigen Rx MMIC IQ Mischers, der die 100 Gbit/s Experimente ermöglicht hat.

Dr. Christian Meuer (TU Berlin, HHI Berlin) und Holger Schmeckebier (TU Berlin) danke ich für zwei unvergessliche Messwochen im IPQ-Systemlabor, die an Effizienz, Gründlichkeit und Spaß wohl kaum zu überbieten sind.

Ich danke meinem Kollegen Dr. René Schmogrow für die Unterstützung bei der Benutzung der Multiformat-Transmitter und für die Diskussionen zu OFDM und DSP.

Meinem Kollegen Dr. Thomas Vallaitis danke ich für die hilfreichen und wertvollen Ratschläge zu Beginn meiner Arbeit und für die Diskussionen über SOAs und Halbleiterphysik.

Meinen Bürokollegen Dr. Moritz Röger, Dr. Jingshi Li, Claudius Weimann, Dr. Stelios Sygletos und Christos Klamouris danke ich für das außerordentlich angenehme Büroklima in all den Jahren und für die freundliche Übernahme des Gießens der Büropflanzen.

Ein großer Dank gilt auch den vielen motivierten und sehr guten Studierenden, von denen jeder seinen Teil zu dieser Arbeit beigetragen hat: Florian Boes, Matthias Hoh, Jörg Pfeifle, Djorn Karnick, Benedikt Bäuerle, Arne Josten, Daniel Lindt, Arnab Bhattacharya, Alejandro Sánchez Santos und George Okyere Dwapanyin.

Ich möchte auch den Mitarbeitern des IPQ danken, die mich auf verschiedenste Weise unterstützt haben. Bei Frau Bernadette Lehmann, Frau Ilse Kober, Frau Angelika Olbrich und Frau Andrea Riemensperger aus dem IPQ-Sekretariat bedanke ich mich für die administrative Unterstützung. Oswald Speck aus dem Packaging-Labor danke ich für seine wertvollen Tipps für das Faserspleißen. Herrn Bürger, Herrn Hirsch und Herrn Höhne aus der mechanischen Werkstatt danke ich für die „kleinen“ mechanischen Arbeiten, ohne die manche Experimente nicht annährend so gut gelungen wären. Herrn Martin Winkeler und Herrn Sebastian Struck danke ich für die Unterstützung im Bereich Elektronik, Computer und Administration.

Danke an alle Kollegen des IPQ für die gute Zusammenarbeit, den Spaß und für die Aktivitäten gelegentlich auch außerhalb des Instituts: Dr. René Bonk, Dr. Moritz Röger, Dr. Thomas Vallaitis, Dr. Jingshi Li, Dr. René Schmogrow, Dr. David Hillerkuss, Prof. Dr. C. Koos, Dr. Sean O’Duill, Dr. Sebastian Köber, Dr. Marcus Winter, Dr. Philipp Vorreau, Dr. Stelios Sygletos, Dr. Luca Alloatti, Dr. Arvind Mishra, Robert Palmer, Jörg Pfeifle, Philipp Schindler, Nicole Lindenmann, Alexandra Ludwig, Dietmar Korn, Christos Klamouris, Argishti Melikyan, Sascha Mühlbrandt, Simon Schneider, Claudius Weimann, Matthias Lauermann, Stefan Wolf, Florian Rupp, Kai Worms und Frans Wegh.

Zu guter Letzt richte ich meinen Dank an das Nachbarinstitut, das Institut für Hochfrequenztechnik und Elektronik (IHE). Ich danke Herrn Prof. Dr.-Ing. Thomas Zwick und allen Kollegen vom IHE für die außerordentliche Gastfreundschaft bei allen IHE-Veranstaltungen, an denen ich als IPQler teilhaben durfte.

List of Publications

Journal Papers

[J1] **S. Koenig**, R. Bonk, H. Schmuck, W. Poehlmann, Th. Pfeiffer, C, Koos, W. Freude, and J. Leuthold, "Amplification of advanced modulation formats with a semiconductor optical amplifier cascade," *Opt. Express*, vol. 22, no. 15, pp. 17854-17871, 2014, doi: 10.1364/OE.22.017854.

[J2] **S. Koenig**, J. Antes, D. Lopez-Diaz, R Schmogrow, T. Zwick, C. Koos, W. Freude, J. Leuthold, and I. Kallfass, "20 Gbit/s Wireless Bridge at 220 GHz," *J. Opt. Commun. Netw.*, vol. 6, no. 1, pp 54-61, 2014, doi: 10.1364/JOCN.6.000054.

[J3] **S. Koenig**, D. Lopez-Diaz, J. Antes, F. Boes, R. Henneberger, A. Leuther, A. Tessmann, R. Schmogrow, D. Hillerkuss, R. Palmer, T. Zwick, C. Koos, W. Freude, O. Ambacher, J. Leuthold, and I. Kallfass, "Wireless sub-THz communication system with high data rate," *Nature Photon.*, vol. 7, no. 12, pp. 977-981, 2013, published online 13 October 2013, doi: 10.1038/nphoton.2013.275.

[J4] R. Palmer, L. Alloatti, D. Korn, P. C. Schindler, R. Schmogrow, W. Heni, **S. Koenig**, J. Bolten, T. Wahlbrink, M. Waldow, H. Yu, W. Bogaerts, P. Verheyen, G. Lepage, M. Pantouvaki, J. Van Campenhout, P. Absil, R. Dinu, W. Freude, C. Koos, and J. Leuthold, "Silicon-Organic Hybrid MZI Modulator Generating OOK, BPSK and 8-ASK Signals for Up to 84 Gbit/s," *IEEE Photon. J.*, vol. 5, no. 2, article no. 6600907, 2013, doi: 10.1109/JPHOT.2013.2258142.

[J5] R. Bonk, G. Huber, T. Vallaitis, **S. Koenig**, R. Schmogrow, D. Hillerkuss, R. Brenot, F. Lelarge, G. H. Duan, S. Sygletos, C. Koos, W. Freude, and J. Leuthold, "Linear semiconductor optical amplifiers for amplification of advanced modulation formats," *Opt. Express*, vol. 20, no. 9, pp. 9657-9672, 2012, doi: 10.1364/OE.20.009657.

[J6] R. Schmogrow, B. Nebendahl, M. Winter, A. Josten, D. Hillerkuss, **S. Koenig**, J. Meyer, M. Dreschmann, M. Huebner, C. Koos, J. Becker, W. Freude, and J. Leuthold, "Error Vector Magnitude as a Performance Measure for Advanced Modulation Formats," *IEEE Photon. Technol. Lett.*, vol. 24, no. 1, pp. 61-63, 2012, doi: 10.1109/LPT.2011.2172405. Correction: ibid., vol. 24, no. 23, p. 2198, 2012, doi: 10.1109/LPT.2012.2219471.

[J7] D. Hillerkuss, R. Schmogrow, T. Schellinger, M. Jordan, M. Winter, G. Huber, T. Vallaitis, R. Bonk, P. Kleinow, F. Frey, M. Roeger, **S. Koenig**, A. Ludwig, A. Marculescu, J. Li, M. Hoh, M. Dreschmann, J. Meyer, S. Ben Ezra, N. Narkiss, B. Nebendahl, F. Parmigiani, P. Petropoulos, B. Resan, A. Oehler, K. Weingarten, T. Ellermeyer, J. Lutz, M. Moeller, M. Huebner, J. Becker, C. Koos, W. Freude, and J. Leuthold, "26 Tbit s^{-1} line-rate super-channel transmission utilizing all-optical fast Fourier transform processing," *Nature Photon.*, vol. 5, no. 6, pp. 364-371, 2011, published online 22 May 2011, doi: 10.1038/nphoton.2011.74.

Conference Contributions

[C1] **S. Koenig**, D. Lopez-Diaz, J. Antes, F. Boes, R. Henneberger, A. Leuther, A. Tessmann, R. Schmogrow, D. Hillerkuss, R. Palmer, T. Zwick, C. Koos, W. Freude, O. Ambacher, J. Leuthold, and I. Kallfass, "Wireless Sub-THz Communication System with High Data Rate Enabled by RF Photonics and Active MMIC Technology," *IEEE Photonics Conference (IPC)*, San Diego (CA), USA, October 2014, invited paper.

[C2] D. Lopez-Diaz, F. Boes, J. Antes, **S. Koenig**, A. Tessmann, F. Kurz, F. Poprawa, R. Henneberger, and I. Kallfass, "Multi-Gigabit Data Transmission at 240 GHz with Complex Baseband Power Detection," *European Microwave Integrated Circuits Conference (EuMIC)*, Rome, Italy, October 2014.

[C3] W. Freude, **S. Koenig**, D. Lopez-Diaz, J. Antes, F. Boes, R. Henneberger, A. Leuther, A. Tessmann, R. Schmogrow, D. Hillerkuss, R. Palmer, T. Zwick, C. Koos, O. Ambacher, J. Leuthold, and I. Kallfass, "Wireless communications on THz carriers takes shape," *16th International Conference on Transparent Optical Networks (ICTON)*, Graz, Austria, July 2014, paper We.D2.1 (invited).

[C4] C. Koos, J. Leuthold, W. Freude, L. R. Dalton, S. Koeber, R. Palmer, C. Weimann, D. L. Elder, W. Heni, D. Korn, J. Pfeifle, S. Wolf, D. Bekele, M. Woessner, L. Alloatti, P.C. Schindler, and **S. Koenig** "Femtojoule modulation and frequency comb generation in silicon-organic hybrid devices," *16th International Conference on Transparent Optical Networks (ICTON)*, Graz, Austria, July 2014, paper We.C2.1 (invited).

[C5] A. Nesic, R. Palmer, S. Koeber, D. Korn, **S. Koenig**, D. L. Elder, L. R. Dalton, W. Freude, and C. Koos, "Demonstration of Difference Frequency Generation in a Silicon Slot Waveguide," *Conference on Lasers and Electro-Optics (CLEO)*, San Jose (CA), USA, June 2014, paper STh1I.2, doi: 10.1364/CLEO_SI.2014.STh1I.2.

[C6] D. Lopez-Diaz, **S. Koenig**, J. Antes, F. Boes, F. Kurz, R. Henneberger, A. Tessmann, A. Leuther, S. Wagner, M. Schlechtweg, O. Ambacher, and I. Kallfass, "A 240 GHz quadrature receiver and transmitter for data transmission up to 40 Gbit/s," *European Microwave Integrated Circuits Conference (EuMIC)*, Nuremberg, Germany, October 2013, pp. 440-443.

[C7] R. Palmer, S. Koeber, W. Heni, D. L. Elder, D. Korn, H. Yu, L. Alloatti, **S. Koenig**, P. C. Schindler, W. Bogaerts, M. Pantouvaki, G. Lepage, P. Verheyen, J. Van Campenhout, P. Absil, R. Baets, L. R. Dalton, W. Freude, J. Leuthold, and C. Koos, "High-speed silicon-organic hybrid (SOH) modulator with 1.6 fJ/bit and 180 pm/V in-device nonlinearity, *39th European Conference and Exhibition on Optical Communication (ECOC'13)*, London, United Kingdom, September 2013, paper We.3.B.3 (invited), doi: 10.1049/cp.2013.1443.

[C8] J. Antes, **S. Koenig**, D. Lopez-Diaz, F. Boes, A. Tessmann, R. Henneberger, O. Ambacher, T. Zwick, and I. Kallfass, "Transmission of an 8-PSK Modulated 30 Gbit/s Signal Using an MMIC-Based 240 GHz Wireless Link," *IEEE International Microwave Symposium (IMS)*, Seattle, USA, June 2013, doi: 10.1109/MWSYM.2013.6697568.

[C9] B. Nebendahl, R. Schmogrow, A. Josten, D. Hillerkuss, **S. Koenig**, M. Winter, J. Meyer, M. Dreschmann, W. Freude, C. Koos, J. Leuthold, and M. Huebner "EVM as new quality metric for optical modulation analysis," *IEEE Saudi International Electronics, Communications and Photonics Conference (SIECPC)*, Riyadh, Saudi Arabia, April 2013, Saudi International, doi: 10.1109/SIECPC.2013.6551002.

[C10] **S. Koenig**, F. Boes, D. Lopez-Diaz, J. Antes, R. Henneberger, R. M. Schmogrow, D. Hillerkuss, R. Palmer, T. Zwick, C. Koos, W. Freude, O. Ambacher, I. Kallfass, and J. Leuthold, "100 Gbit/s Wireless Link with mm-Wave Photonics," *Optical Fiber Communication Conference (OFC)*, Anaheim (CA), USA, March 2013, postdeadline paper PDP5B.4., doi: 10.1364/NFOEC.2013.PDP5B.4.

[C11] **S. Koenig**, R. Bonk, R. M. Schmogrow, A. Josten, D. Karnick, H. Schmuck, W. Poehlmann, Th. Pfeiffer, C. Koos, W. Freude, and J. Leuthold, "Cascade of 4 SOAs with 448 Gbit/s (224 Gbit/s) Dual Channel Dual Polarization 16QAM (QPSK) for High-Capacity Business Paths in Converged Metro-Access Networks," *Optical Fiber Communication Conference (OFC)*, Anaheim (CA), USA, March 2013, paper OTh4A.3, doi: 10.1364/OFC.2013.OTh4A.3.

[C12] R. Palmer, L. Alloatti, D. Korn, P. C. Schindler, R. M. Schmogrow, M. Baier, **S. Koenig**, D. Hillerkuss, J. Bolten, T. Wahlbrink, M. Waldow, R. Dinu, W. Freude, C. Koos, and J. Leuthold, "Silicon-Organic Hybrid (SOH) Modulator Generating up to 84 Gbit/s BPSK and M-ASK Signals," *Optical Fiber Communication Conference (OFC)*, Anaheim (CA), USA, March 2013, paper OW4J.6, doi: 10.1364/OFC.2013.OW4J.6.

[C13] B. Nebendahl, R. Schmogrow, T. Dennis, A. Josten, D. Hillerkuss, **S. Koenig**, J. Meyer, M. Dreschmann, M. Winter, M. Huebener, W. Freude, C. Koos, and J. Leuthold, "Quality Metrics in Optical Modulation Analysis: EVM and its relation to Q-factor, OSNR, and BER," *Asia Communications and Photonics Conference (APC)*, Guangzhou, China, November 2012, paper AF3G.2.

[C14] W. Freude, R. Schmogrow, B. Nebendahl, M. Winter, A. Josten, D. Hillerkuss, **S. Koenig**, J. Meyer, M. Dreschmann, M. Huebner, C. Koos, and J. Leuthold, "Quality metrics for optical transmission," *International Conference on Photonics in Switching (PS)*, Ajaccio (Corsica), France, September 2012, (invited).

[C15] W. Freude, R. Schmogrow, B. Nebendahl, M. Winter, A. Josten, D. Hillerkuss, **S. Koenig**, J. Meyer, M. Dreschmann, M. Huebner, C. Koos, J. Becker, and J. Leuthold, "Quality metrics for optical signals: Eye diagram, Q-factor, OSNR, EVM and BER," *14th International Conference on Transparent Optical Networks (ICTON)*, Coventry, United Kingdom, July 2012, paper Mo.B1.5 (invited), doi: 10.1109/ICTON.2012.6254380.

[C16] A. K. Mishra, Z. Wang, H. Klein, R. Bonk, **S. Koenig**, D. Karnick, R. Schmogrow, D. Hillerkuss, M. Moehrle, T. Pfeiffer, C. Koos, J. Leuthold, and W. Freude, "Performance analysis of an OFDM transmission system with directly modulated lasers for wireless backhauling," *14th International Conference on Transparent Optical Networks (ICTON)*, Coventry, United Kingdom, July 2012, Mo.C3.3 (invited), doi: 10.1109/ICTON.2012.6254423.

[C17] J. Antes, **S. Koenig**, A. Leuther, H. Massler, J. Leuthold, O. Ambacher, and I. Kallfass, "220 GHz wireless data transmission experiments up to 30 Gbit/s," *IEEE International Microwave Symposium (IMS)*, Montreal, Canada, June 2012, doi: 10.1109/MWSYM.2012.6259561.

[C18] **S. Koenig**, J. Antes, D. Lopez-Diaz, I. Kallfass, T. Zwick, C. Koos, W. Freude, and J. Leuthold, "High-speed wireless bridge at 220 GHz connecting two fiber-optic links each spanning up to 20 km," *Optical Fiber Communication Conference (OFC)*, Los Angeles (CA), USA, March 2012, paper OM2B.1, doi: 10.1364/OFC.2012.OM2B.1.

[C19] J. Leuthold, W. Freude, C. Koos, R. Bonk, **S. Koenig**, D. Hillerkuss, and R. Schmogrow, "Semiconductor Optical Amplifiers in Extended Reach PONs," *Access Networks and Inhouse Communications (ANIC)*, Toronto, Canada, June 2011, paper ATuA1 (invited), doi: 10.1364/ANIC.2011.ATuA1.

[C20] S. Mumtaz, J. Li, **S. Koenig**, Y. Jaouen, R. Schmogrow, G. Rekaya-Ben Othman, and J. Leuthold, "Experimental Demonstration of PDL Mitigation using Polarization-Time Coding in PDM-OFDM Systems," *Signal Processing in Photonic Communication (SPPCom)*, Toronto, Canada, June 2011, paper SPWB6, doi: 10.1364/SPPCOM.2011.SPWB6.

[C21] S. Mumtaz, G. Rekaya-Ben Othman, Y. Jaouen, J. Li, **S. Koenig**, R. Schmogrow, and J. Leuthold, "Alamouti Code against PDL in Polarization Multiplexed Systems," *Signal Processing in Photonic Communications (SPPCom)*, Toronto, Canada, June 2011, paper SPTuA2, doi: 10.1364/SPPCOM.2011.SPTuA2.

[C22] J. Antes, D. Lopez-Diaz, M. Schlechtweg, S. Diebold, H. Gulan, F. Poprawa, F. Kurz, R. Henneberger, E. Roppelt, **S. Koenig**, J. Leuthold, T. Schneider, T. Zwick, O. Ambacher, and I. Kallfass, "Systemkonzept und Realisierung einer Millimeterwellen-Richtfunkstrecke mit Datenraten über 12,5 Gbit/s," *16. ITG-Fachtagung Mobilkommunikation-Technologien und Anwendungen*, Osnabrück, Germany, Mai 2011.

[C23] **S. Koenig**, M. Hoh, R. Bonk, H. Wang, P. Pahl, T. Zwick, C. Koos, W. Freude, and J. Leuthold, "Rival signals in SOA reach-extended WDM-TDM-GPON converged with RoF," *Optical Fiber Communication Conference (OFC)*, Los Angeles (CA), USA, March 2011, paper OWT1, doi: 10.1364/OFC.2011.OWT1.

[C24] M. Roeger, F. Boes, A. Kleff, B. Hiba, M. Baier, M. Hoh, **S. Koenig**, C. Koos, J. Leuthold, and W. Freude, "Energy-efficient MAC protocol enabling an optically powered sensor network," *Optical Fiber Communication Conference (OFC)*, Los Angeles (CA), USA, March 2011, poster JWA88, doi: 10.1364/NFOEC.2011.JWA088.

[C25] **S. Koenig**, J. Pfeifle, R. Bonk, T. Vallaitis, C. Meuer, D. Bimberg, C. Koos, W. Freude, and J. Leuthold, "Optical and electrical power dynamic range of semiconductor optical amplifiers in radio-over-fiber networks," *36th European Conference and Exhibition on Optical Communication (ECOC'10)*, Torino, Italy, September 2010, paper Th.10.B.6, doi: 10.1109/ECOC.2010.5621535.

Advised Students' Theses

[S1] Florian Boes, "100 Gbit/s mm-Wave Photonic Wireless Link," Master's thesis, *Karlsruhe Institute of Technology (KIT), Institute of Photonics and Quantum Electronics (IPQ)*, February 2013.

[S2] Djorn Karnick, "Linear Semiconductor Optical Amplifiers in Reconfigurable Optical Add/Drop Multiplexers for Next Generation Optical Metro/Access Networks," Diplomarbeit, *Karlsruhe Institute of Technology (KIT), Institute of Photonics and Quantum Electronics (IPQ)*, February 2012.

[S3] Daniel Lindt, "Semiconductor Optical Amplifiers in ROADMs for future Converged Metro/Access Ring Networks," Bachelor's thesis, *Karlsruhe Institute of Technology (KIT), Institute of Photonics and Quantum Electronics (IPQ)*, December 2011.

[S4] Matthias Hoh, "Rival Signals in Semiconductor Optical Amplifier Reach-Extended WDM-TDM-Gigabit Passive-Optical Network Converged with Radio-over-Fiber," Diplomarbeit, Karlsruhe Institute of Technology (KIT), Institute of Photonics and Quantum Electronics (IPQ), February 2011.

[S5] Djorn Karnick, "Directly Modulated Lasers for OFDM," Studienarbeit, *Karlsruhe Institute of Technology (KIT), Institute of Photonics and Quantum Electronics (IPQ)*, December 2010.

[S6] Arne Josten, "OFDM in Future Access Networks," Bachelor's thesis, *Karlsruhe Institute of Technology (KIT), Institute of Photonics and Quantum Electronics (IPQ)*, November 2010.

[S7] Benedikt Bäuerle, "Generation of Electrical OFDM for Future Access Networks," Bachelor's thesis, *Karlsruhe Institute of Technology (KIT), Institute of Photonics and Quantum Electronics (IPQ)*, October 2010.

[S8] Jörg Pfeifle, "Semiconductor Optical Amplifiers in Radio-over-Fiber Networks," Diplomarbeit, *Karlsruhe Institute of Technology (KIT), Institute of Photonics and Quantum Electronics (IPQ)*, May 2010.

[S9] Alejandro Sánchez Santos, "Simulation and Performance Analysis of Multimode Fibers for Radio-over-Fiber Systems," Diplomarbeit, *Karlsruhe Institute of Technology (KIT), Institute of Photonics and Quantum Electronics (IPQ)*, November 2009.

[S10] Arnab Bhattacharya, "Setup of a Radio-over-Fiber Link and Study of the Inclusion of a Quantum Dot Semiconductor Optical Amplifier to Enhance the Link Performance," Studienarbeit, *Karlsruhe Institute of Technology (KIT), Institute of Photonics and Quantum Electronics (IPQ)*, July 2009.

Curriculum Vitae

Swen König
Swen.Koenig@web.de

Work Experience

06/2008 – 06/2013 **Karlsruhe Institute of Technology (KIT), Karlsruhe, Germany, Institute of Photonics and Quantum Electronics**

PhD studies

- Semiconductor optical amplifiers (SOA) and cascaded SOAs for optical access networks.
- Radio-over-fiber (RoF), μ-wave and mm-wave photonics.
- Research within the framework of European and national projects with focus on metro-access network convergence and optical-wireless convergence.
- Contribution to work packages and deliverables of several European and national research projects
- Teaching assistant tutorial lecturer for lectures on semiconductor devices, optoelectronic components, optical waveguides, senders, and receivers.
- Advisor of BSc and MSc student projects.

07/2002 – 09/2002 **Siemens AG, Stuttgart, Germany**
Internship
Metal processing and electronic circuit design.

05/2002 – 06/2002 **Porsche AG, R&D Center, Weissach, Germany**
Internship
Testing of automotive electronics and sensors.

Education

06/2008 – 07/2013	**Karlsruhe Institute of Technology (KIT), Karlsruhe, Germany, Institute of Photonics and Quantum Electronics** PhD thesis on "Semiconductor Optical Amplifiers and mm-Wave Wireless Links for Converged Access Networks" Defense on July 8, 2013. Summa cum laude ("mit Auszeichnung").
10/2002 – 03/2008	**University of Stuttgart, Stuttgart, Germany** Dipl.-Ing. in Electrical Engineering and Information Technology Specialization micro- and optoelectronics
09/2006 – 03/2008	**Télécom ParisTech (ENST), Paris, France** (German-French double degree program) Diplôme d'ingénieur (M.Eng.) in Communications Engineering Specialization optical and wireless communications
07/2007– 03/2008	**Joint Lab CEA-LETI & STMicroelectronics, Grenoble, France** Diplomarbeit / Stage d'Ingénieur on "Performance Analysis of Phased Array Architectures at mm-Wave Frequencies"
07/2001– 05/2002	**Alternative civilian national service (Zivildienst)**
08/1992– 06/2001	**Max Planck Gymnasium Schorndorf, Germany** Abitur (university entrance qualification)